Manfred Denker

Asymptotic Distribution Theory in Nonparametric Statistics

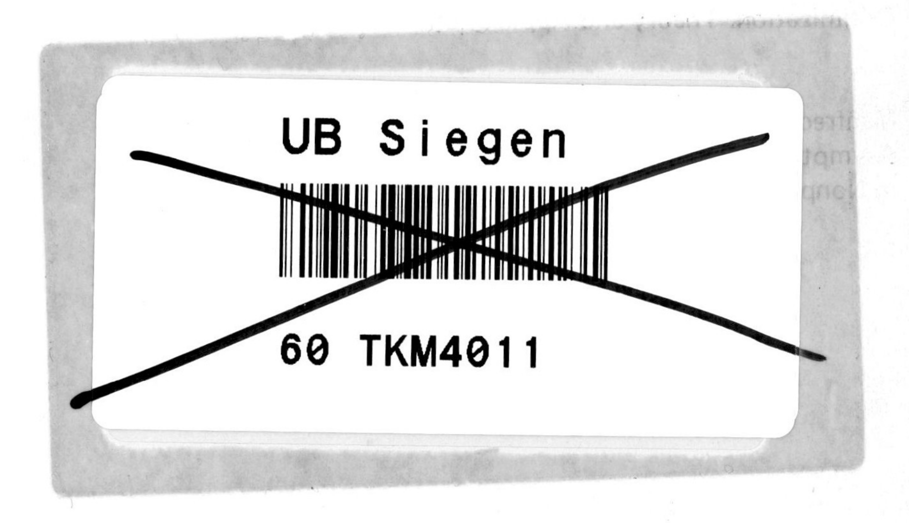

Advanced Lectures in Mathematics

Edited by Gerd Fischer

Jochen Werner
Optimization. Theory and Applications

Manfred Denker
Asymptotic Distribution Theory
in Nonparametric Statistics

Manfred Denker

Asymptotic Distribution Theory in Nonparametric Statistics

Friedr. Vieweg & Sohn Braunschweig/Wiesbaden

CIP-Kurztitelaufnahme der Deutschen Bibliothek

Denker, Manfred:
Asymptotic distribution theory in nonparametric statistics / Manfred Denker.– Braunschweig; Wiesbaden: Vieweg, 1985.
(Advanced lectures in mathematics)
ISBN 3-528-08905-9

1985

Produced by IVD, Walluf b. Wiesbaden
Printed in Germany

ISBN 3-528-08905-9

PREFACE

During the last years I gave some courses at the University of Göttingen on selected topics in nonparametric statistics, mainly on its asymptotic distribution theory. They were intended for advanced students with a good mathematical background, especially in probability theory and mathematical statistics. The present notes result from these lectures.
Three basic types of statistics are treated in this book: Chapter 1 contains Hoeffding's U-statistics, chapter 2 differentiable statistical functionals and chapter 3 statistics based on ranks. Although the emphasis lies on the asymptotic distribution theory for these statistics I intended to give some motivation from two viewpoints. The examples cover more than is needed for illustrations and each chapter contains some result on optimality properties. Chapter 4 on efficiency and contiguity may be regarded as an application of the results in the preceding chapters.
The results are presented with complete proofs. For this reason a good deal of probability is needed for parts of the proofs. Billingsley's book (1968) is an adequate reference for it. For example, Donsker's theorem on the weak convergence of the empirical process to the Brownian bridge is the essential point for the development of the asymptotic distribution theory for von Mises' statistics in chapter 2, section 2. For most of the results in the other sections, however, a basic knowledge in probability suffices, like the central limit theorem under the Lindeberg condition.
I intended to restrict the content to results which can be proved in a reasonable way during a course, even concerning its technical part. Thus some of the theorems do not appear in full generality and very often, the assumptions are not as weak as possible. Unfortunately, the asymptotic distribution theory in nonparametric statistics requires some rather extensive computations. The techniques for doing this I tried to develop carefully for each of the statistics under consideration. Some of the material in this book has not yet appeared

in print elsewhere and therefore I hope that these notes are helpful not only for students to get acquainted with some theoretical aspects of nonparametric asymptotic distribution theory.

Certainly, in a volume like this, I had to make choices of the material to be included and the reader might miss essential and important other classes of statistics, e.g. Kolmogorov-Smirnov statistics, statistics based on empirical processes in general or statistics treated by martingale methods, to name a few. In recent years some books were published containing more and different matters about the mathematical theory in nonparametric statistics, on the other hand these notes go beyond previous presentations.

I am very much indebted to Chr. Grillenberger, G. Keller and U. Rösler, whose valuable collaborations resulted in joint papers and made this book possible. Especially, I have to thank H. Dehling and Chr. Grillenberger for helpful comments while reading parts of the manuscript, and also M. Powell for assisting in typing. Also, I am grateful to the Vieweg Verlag for publishing the manuscript.

Göttingen, September 1984

Contents

CHAPTER 1: U-STATISTICS

Hoeffding introduced U-statistics in 1948, partly influenced by earlier work of Halmos, and closely connected to von Mises' functionals (von Mises, 1947). U-statistics can be viewed as a class of unbiased estimators of a certain parameter, based on some averaging procedure.

We shall investigate the asymptotic properties of non-degenerate U-statistics in the next three sections; section 4 gives extensions to Lehmann's generalized U-statistics. More results on U-statistics will appear in the next chapters, especially the degenerate case in chapter 2 and contiguity in chapter 4.

1. Definition of U-statistics

For a measurable space $(E,\mathcal{B})$ denote by $M(\mathcal{B})$ the set of all probability measures on $\mathcal{B}$.

Definition 1.1.1: A function $\vartheta : M_o \to \mathbb{R}$, defined on a subset $M_o \subset M(\mathcal{B})$, is said to be *regular* (*estimable*) w.r.t. M_o, if there exist an integer $m \geq 1$ and an unbiased estimator h based on m independent, identically distributed (= i.i.d.), E-valued random variables.

This definition means that there is a measurable map $h : E^m \to \mathbb{R}$ such that

$$E_{P^m} h = \int h \, dP^m = \vartheta(P) \qquad (\forall\ P \in M_o),$$

where P^m is the m-fold Cartesian product of P . If ϑ is given, the minimum of all integers m with this property is called the *rank* of ϑ. It should be remarked that h can always be assumed to be symmetric. Indeed, if γ_m denotes the permutation group acting on E^m by permuting the coordinates, then $(m!)^{-1} \sum_{\tau \in \gamma_m} h \circ \tau$ is an unbiased symmetric estimator for ϑ (assuming that h is unbiased).

The definition of a U-statistic is based on this notion of regular functions ϑ .

Definition 1.1.2: Let $\vartheta : M_o \to \mathbb{R}$ be regular. If $h : E^m \to \mathbb{R}$ is a symmetric, unbiased estimator for ϑ, then for any $n \geq m$ the map

$$U_n(h) = \binom{n}{m}^{-1} \sum_{\pi \in \Pi_n} h \circ \pi$$

is called a *U-statistic*. Here, Π_n denotes the set of all maps $\pi : E^n \to E^m$ which arise from projections onto m coordinates $1 \leq i_1 < i_2 < \ldots < i_m \leq n$. The estimator h is called the *kernel* of the U-statistic and m is called its degree.

Note that $U_n(h)$ is again an unbiased, symmetric estimator for ϑ based on n i.i.d. observations $X_1, \ldots, X_n$ with distribution $P \in M_o$. Then the U-statistic can be written as

$$U_n(h) = \binom{n}{m}^{-1} \sum_{1 \leq i_1 < \ldots < i_m \leq n} h(X_{i_1}, \ldots, X_{i_m});$$

for convenience, the dependence on the X_i's is suppressed in $U_n(h)$.

Example 1.1.1: Let $f : E \to \mathbb{R}$ be a measurable function and $M_o = \{P \in M(\mathcal{B}) : f \in L_1(P)\}$. Then $\vartheta(P) = \int f \, dP$ is regular (of rank 1) and $h = f$ is an unbiased estimator of rank 1. For $f(x) = x^k$ $(E = \mathbb{R})$, one has the k-th moment and the corresponding U-statistic $1/n \sum_{i=1}^{n} X_i^k$ is known as the *sample k-th moment* (for $k=1$ this is the sample mean).

If $E = \mathbb{R}$, $f(x) = 1_{(-\infty, t]}(x)$ $(t \in \mathbb{R})$ estimates $F(t)$, where F denotes the right continuous distribution function (= d.f.) of X_1. The corresponding U-statistic is just the *empirical d.f.* of $X_1, \ldots, X_n$ evaluated at t.

Example 1.1.2: Let $E = \mathbb{R}$ and $M_o = \{P : \int x^2 dP(x) < \infty\}$. In order to estimate $\vartheta(P) = \int x^2 \, dP(x) - (\int x \, dP(x))^2$, the variance of P, one can use the estimator $h(x,y) = 1/2 \, (x-y)^2$ of rank 2. The corresponding U-statistic gives the *sample variance*

$$U_n(h) = \frac{1}{n(n-1)} \sum_{1 \leq i < j \leq n} (X_i - X_j)^2 = \frac{1}{n-1} \left[\sum_{i=1}^{n} X_i^2 - \left(\sum_{j=1}^{n} X_j\right)^2 \right].$$

ϑ can not be estimated by a rank 1 estimator unless M_o is restricted considerably.

U-statistics, when used for estimation, have a very nice optimality property. This will be discussed first. For real valued random variables $X_1,\ldots,X_s$ denote the *order statistic* by $X^{(s)} = (X_{(1)},\ldots,X_{(s)})$ (i.e., $X_{(1)} \leq X_{(2)} \leq\ldots\leq X_{(s)}$), and denote by $\mathcal{F}_s$ the σ-algebra generated by $X^{(s)}, X_{s+1},\ldots$.

Lemma 1.1.1: Let $U_n(h)$ $(n\geq m)$ be a sequence of U-statistics based on i.i.d. real-valued random variables $(X_k : k \geq 1)$. Then

(1) $\mathcal{F}_{n+1} \subset \mathcal{F}_n$ $(n\geq m)$.

(2) $U_n(h)$ is $\mathcal{F}_n$ measurable $(n \geq m)$.

(3) $U_n(h) = E(U_l | \mathcal{F}_n)$ $m\leq l\leq n$.

Proof: (1) and (2) are evident. In order to show (3) denote by $\hat{E}(\cdot|\cdot)$ the conditional expectation with respect to the image measure on E^n under the map $X^n = (X_1,\ldots,X_n)$, and denote by $\mathcal{B}_n \subset \mathcal{B}^n$ the sub-σ-algebra of all γ_n-invariant sets. If $\pi \in \Pi_n$ is fixed choose $\tau^{-1} \in \gamma_n$ mapping the coordinates $i_1<\ldots<i_m$ defining π into $1,\ldots,m$. Then

$$E(h \circ \pi \circ X^n | \mathcal{F}_n) = \hat{E}(h \circ \pi | \mathcal{B}_n) = \hat{E}(h \circ \pi \circ \tau \mid \tau \mathcal{B}_n)$$
$$= E(h \circ X^m \mid \mathcal{F}_n).$$

Hence by (2)

$$U_n(h) = E(U_n(h)| \mathcal{F}_n) = \binom{n}{m}^{-1} \sum_{\pi\in\Pi_n} E(h \circ \pi \circ X^n \mid \mathcal{F}_n)$$
$$= E(h \circ X^m \mid \mathcal{F}_n) = E(U_m(h) \mid \mathcal{F}_n),$$

and (3) follows.

Remark: (2) and (3) together just mean that the sequence $U_n(h)$ is a reversed martingale. (When speaking of a U-statistic with kernel h it is always assumed by definition that h is properly integrable with respect to the distributions under consideration!). In particular, a U-statistic can be represented by

$$U_n(h) = E(U_m(h) | \mathcal{F}_n).$$

Since the order statistic is sufficient (w.r.t. all permutation-invariant probabilities) the following theorem is in fact a special case of the Rao-Blackwell theorem.

Theorem 1.1.1: For any unbiased estimator S of ϑ there exists a U-statistic U based on the same n observations such that

$$\text{Var } U \leq \text{Var } S .$$

Equality holds if and only if $U = S$ a.s.
If, moreover, M_o consists of a complete family of probability measures, then the U-statistic

$$U = (n!)^{-1} \sum_{\tau \in \Upsilon_n} S \circ \tau$$

constitutes the minimum variance unbiased estimator for ϑ .

Proof: Since the order statistic is sufficient, the second part of the theorem is a special case of the theorem of Lehmann and Scheffé. On the other hand, here is a quick proof of the first part. Let $S = S(X_1,\ldots,X_n)$ and define

$$U = (n!)^{-1} \sum_{\tau \in \Upsilon_n} S \circ \tau(X_1,\ldots,X_n).$$

Note that the right-hand side is equal to $E(S | X^{(n)})$ so that

$$EU^2 = E(E(S | X^{(n)}))^2 \leq E(E(S^2 | X^{(n)})) = ES^2$$

with equality iff $U = S$. Since, moreover, $EU = ES$ the first part of the theorem follows.

An easy application of Lemma 1.1.1 together with a convergence theorem for reversed martingales gives

Theorem 1.1.2: Let $(U_n(h) : n \geq m)$ be a sequence of U-statistics. Then

$$U_n \to \vartheta$$

a.s. and in L_1 as $n \to \infty$.

<u>Proof:</u> Let A_n be a decreasing sequence of σ-algebras and let Z be integrable. Then

$$\lim_{n\to\infty} E(Z|A_n) = E(Z|\bigcap_n A_n) \quad \text{a.s. and in } L_1.$$

By Lemma 1.1.1. $U_n(h) = E(U_m(h)|\mathcal{F}_n)$, hence

$$\lim U_n(h) = E(U_m(h)|\bigcap_{n\geq m} \mathcal{F}_n)$$

a.s. and in L_1. But since $\bigcap \mathcal{F}_n$ is permutation-invariant, the 0-1-law of Hewitt and Savage applies and so $\bigcap_{n\geq m} \mathcal{F}_n$ is trivial, equivalently

$$E(U_m(h)\;|\bigcap \mathcal{F}_n) = E\; U_m(h) = \vartheta.$$

This last theorem especially ensures that the variance of a distribution (if it exists) is consistently estimated by the sample variance in Example 1.1.2.

Quite a few examples of U-statistics arise in testing situations, some of them will be mentioned briefly. Because of their easy structure and their nice mathematical properties they are used frequently.

<u>Example 1.1.3:</u> Let F be a continuous d.f. which is symmetric about some $\vartheta \in \mathbb{R}$. (The test problem is specified by the hypothesis $H_o = \{F : F$ continuous and symmetric about $\vartheta_o\}$ (ϑ_o known) against all other symmetric, continuous d.f. . This can be done using the *Wilcoxon signed rank statistik* W^+.) Let $X_1,\ldots,X_n$ be i.i.d. random variables with d.f. F. Denote by R_i^+ the rank of $|X_i-\vartheta_o|$ among all absolute values $|X_1-\vartheta_o|,\ldots,|X_n-\vartheta_o|$. Then

$$W^+ = 1/2 \sum_{i=1}^{n} R_i^+(1+\text{sign}(X_i-\vartheta_o)).$$

W^+ can be written as a sum of two U-statistics with kernels $h_1(x) = 1_{(0,\infty)}(x)$ and $h_2(x,y) = h_1(x+y)$, based on the sample

$X_1-\vartheta_o,\ldots,X_n-\vartheta_o$, i.e. (exercise!)

$$W^+ = n\,U_n(h_1) + \binom{n}{2}U_n(h_2).$$

From this representation it follows that

$$EW^+ = n\,P(X_1>\vartheta_o) + \binom{n}{2}P(X_1+X_2>2\vartheta_o)$$

and that under H_o

$$EW^+ = \frac{n(n+1)}{4}.$$

The U-statistic corresponding to the kernel $1-h_2$ is called the *Wilcoxon one-sample statistic*.

For small sample sizes the distribution of W^+ under H_o is explicitly computable; however, for large sample sizes ($n>20$) the approximating normal distribution derived from Theorem 1.3.1 below is good enough and much easier to handle.

Example 1.1.4: There is a very interesting way of testing for exponential distribution, proposed by Hollander and Proschan (1972). Let $X_1,\ldots,X_n$ be positive i.i.d. random variables with continuous d.f. F on $\mathbb{R}_+$. One might test for exponential F using the parameter

$$\vartheta = \int_0^\infty\int_0^\infty (1-F(s+t))\,dF(s)\,dF(t).$$

Indeed, (exercise!) if F satisfies

$$1 - F(s+t) = (1-F(s))(1-F(t))$$

then it is exponential. Hence, under H_o,

$$\vartheta = \left(\int_0^\infty (1-F(t))\,dF(t)\right)^2 = 1/4.$$

Define $\hat{h}(x,y,z) = 1_{\{x-y-z>0\}}$ so that

$$\vartheta = \iiint \hat{h}(x,y,z)\,dF(x)\,dF(y)\,dF(z).$$

Setting $h(x,y,z) = 1/3\,(\hat{h}(x,y,z) + \hat{h}(y,x,z) + \hat{h}(z,x,y))$ one obtains a symmetric, unbiased estimator for ϑ and the corresponding U-statistic can be used to test for exponential distribution.

Example 1.1.5: When the underlying distribution is known to be symmetric, the Wilcoxon signed rank test is preferable to the sign test. Hence it is desirable in certain cases to decide for symmetry. Let $X_1,\dots,X_n$ be i.i.d. real-valued random variables with continuous d.f. F. The test problem is given by the hypothesis $H_o = \{F : F \text{ continuous, symmetric}\}$ against all other continuous d.f. .

Consider three values $a<b<c$ and look at all three possibilities: $c-b \gtreqless b-a$. If F is symmetric the first situation should occur approximately as often as the third one when choosing three different values out of $X_1,\dots,X_n$. This heuristic approach leads to the triple test:

Set $f(t) = \begin{cases} 1 & t>0 \\ 0 & t=0 \\ -1 & t<0 \end{cases}$ and

$h(x,y,z) = 1/3[f(x+y-2z)+f(x+z-2y)+f(y+z-2x)]$.

The corresponding U-statistic will then give the test statistic. When estimating its limit variance consistently, the statistic will be asymptotically distribution-free. Also note that

$$\vartheta = Eh = P(X_1+X_2>2X_3) - P(X_1+X_2<2X_3)$$

and that under H_o $X_1+X_2-2X_3$ is an odd, translation invariant statistic and therefore $\vartheta = 0$.

Example 1.1.6: Let $X_1,\dots,X_n$ be i.i.d. $\mathbb{R}^2$-valued random variables with d.f. $F((x,y))$.

$$\vartheta(F) := \iint[F((x,y))-F((x,\infty))F((\infty,y))]^2 dF(x,y)$$

is a measure of dependence between the marginals of F. Let $f(x,y,z) = 1_{\{y\leq x\}} - 1_{\{z\leq x\}}$ and

$h((x_1,y_1),\dots,(x_5,y_5)) =$

$= 1/4\ f(x_1,x_2,x_3)f(x_1,x_4,x_5)f(y_1,y_2,y_3)f(y_1,y_4,y_5)$.

The corresponding U-statistic estimates ϑ !

Example 1.1.7: Let P be a known distribution, $A_i \in \mathcal{B}$ $(1 \leq i \leq l)$ a partition of E and let $X_1,\ldots,X_n$ be i.i.d. random variables. In order to test whether the distribution of X_1 is P one can use the χ^2-test of fit. Let n_i be the number of all X_j's falling into A_i and let $p_i = P(A_i)$. Then

$$\chi^2 = \sum_{i=1}^{l} \frac{(n_i - np_i)^2}{np_i}$$

is called the χ^2-*statistic* (for the χ^2-test of fit). Since $n_i = \sum_{j=1}^{n} 1_{A_i}(X_j)$, χ^2 can be rewritten as

$$\chi^2 = \frac{1}{n} \sum_{k=1}^{n} \sum_{j=1}^{n} \sum_{i=1}^{l} p_i^{-1} (1_{A_i}(X_k) - p_i)(1_{A_i}(X_j) - p_i).$$

This is almost a U-statistic with kernel

$$h(x,y) = \sum_{i=1}^{l} p_i^{-1} (1_{A_i}(x) - p_i)(1_{A_i}(y) - p_i).$$

Asymptotically, $\frac{1}{n}\chi^2$ is equivalent to the U-statistic U_n with kernel h. Note that $E\frac{1}{n}\chi^2 \neq E\,U_n$ in general. Under H_o, $Eh = 0$ and hence $E\,U_n = 0$, while $E\frac{1}{n}\chi^2 = \frac{1}{n}\sum_{i=1}^{l}(1-p_i)^2$. Both statistics can be used as test of fit.

This last example motivates the following two definitions:

Definition 1.1.3: Let $(E,\mathcal{B})$ be a measurable space and let $h : E^m \longrightarrow \mathbb{R}$ be measurable and symmetric. A *V-statistic* is defined to be a map $V_n : E^n \longrightarrow \mathbb{R}$ of the form

$$V_n = V_n(h) = n^{-m} \sum_{i_1=1}^{n} \cdots \sum_{i_m=1}^{n} h(x_{i_1},\ldots,x_{i_m}).$$

If $m=2$ then $V_n = \frac{n-1}{n} U_n + \frac{1}{n^2} \sum_{i=1}^{n} h(X_i,X_i)$, and hence a V-statistic never is unbiased unless $E\,h(X_1,X_1) = E\,h(X_1,X_2)$. In the last example, under H_o, the kernel h satisfies

$$\int h(x,y)\, dF(x) = 0 \quad \text{for all } y.$$

Thus h is degenerate according to

Definition 1.1.4: Let $h : E^m \to \mathbb{R}$ be a symmetric unbiased estimator for $\vartheta(P)$ $(P \in M_o)$. h is said to be *degenerate at* $P \in M_o$ if for some $m_o < m$

$$\int \cdots \int h(x_1,\ldots,x_m) \prod_{i=1}^{m_o} dP(x_i) = 0$$

for all $x_{m_o+1},\ldots,x_m$. Otherwise h is said to be *non-degenerate*.

We shall use the notion of degeneracy only in the case when $m_o = 1$.

2. The decomposition theorem for a U-statistic

In this section we discuss Hoeffding's analytical method to investigate asymptotic properties of U-statistics. It is based on his decomposition theorem for U-statistics, often also called the projection method.

To begin with we will need some more notation. Let $h : E^m \to \mathbb{R}$ be a (symmetric) kernel of degree m and fix a probability Q on E satisfying $\int \cdots \int |h(x_1,\ldots,x_m)| \prod_{i=1}^{n} dQ(x_i) < \infty$ (i.e. $Q \in M_o$). For $0 \leq c \leq m$ denote by $\mathcal{B}^c$ the σ-algebra generated by the first c coordinates. Then the conditional expectation $E(h|\mathcal{B}^c)$ of h w.r. to the measure Q^m is a.s. independent of the coordinates $x_{c+1},\ldots,x_m$ and hence

$$(1.2.1)\quad \tilde{h}_c(x_1,\ldots,x_c) = E(h|\mathcal{B}^c)(x_1,\ldots,x_c) \quad Q^m\text{ - a.s.}$$

is a well-defined function. If $X_1,\ldots,X_m$ are i.i.d. with distribution Q , setting $X^j := (X_1,\ldots,X_j)$ $(0 \leq j \leq m)$, it follows immediately that

$$(1.2.2)\quad \begin{cases} \tilde{h}_c(x_1,\ldots,x_c) = \mathbf{E}\, h(x_1,\ldots,x_c, X_{c+1},\ldots,X_m) \\ \tilde{h}_c(X_1,\ldots,X_c) = E(h \circ X^m \mid X^c) \\ \tilde{h}_c(x_1,\ldots,x_c) = \int \cdots \int h(x_1,\ldots,x_m) \prod_{i=c+1}^{m} dQ(x_i) \\ \tilde{h}_c(x_1,\ldots,x_c) = \int \tilde{h}_{c+1}(x_1,\ldots,x_{c+1})\, dQ(x_{c+1}) \\ \tilde{h}_o \equiv \vartheta,\ \tilde{h}_m = h. \end{cases}$$

If $x = (x_1,\ldots,x_c) \in E^c$ and if $J = \{c_1 < \ldots < c_j\} \subset \{1,\ldots,c\}$ let x_J denote the point in E^j with coordinates $x_{c_1},\ldots,x_{c_j}$. With these notations define the function $h_c : E^c \to \mathbb{R}$ by

$$(1.2.3)\quad h_c(x) := \sum_{k=0}^{c} (-1)^{c-k} \sum_{\substack{J\subset\{1,..,c\}\\ |J|=k}} \tilde{h}_k(x_J).$$

Lemma 1.2.1:

(1) $h_0 \equiv \vartheta$

(2) Each $\tilde{h}_c$ and each h_c is symmetric in its arguments.

(3) Each h_c (for $0 < c \leq m$) is degenerate, that is

$$\int h_c(x_1,\ldots,x_c)\,dQ(x_i) = 0$$

for all $1 \leq i \leq c$ and all $x_1,\ldots,x_c \in E$.

(4) $$h(x_1,\ldots,x_m) = \sum_{c=0}^{m} \sum_{\substack{K\subset\{1,..,m\}\\ |K|=c}} h_c(x_K).$$

Proof:

(1) is obvious.

(2) Each $\tilde{h}_k$ clearly is symmetric. To see this also for h_c, let $1 \leq i < j \leq c$ and $J \subset \{1,\ldots,c\}$. If y denotes the point derived from x by interchanging the coordinates i and j, then clearly $\tilde{h}_k(x_J) = \tilde{h}_k(y_J)$ ($|J|=k$) if $i,j \in J$ or if $i,j \notin J$. If $i \in J$ and $j \notin J$ then one can find in a unique way some $K \subset \{1,\ldots,c\}$ with $|K| = |J|$, $i \notin K$, $j \in K$ and $\tilde{h}_k(x_J) = \tilde{h}_k(y_K)$. This shows the symmetry using the definition of h_c in (1.2.3).

(3) Fix $c > 0$ and $1 \leq i \leq c$. Then

$$\int h_c(x_1,\ldots,x_c)\,dQ(x_i) =$$

$$= \sum_{k=0}^{c} (-1)^{c-k} \sum_{\substack{|J|=k\\ i\notin J}} \tilde{h}_k(x_J) + \sum_{k=1}^{c} (-1)^{c-k} \sum_{\substack{|J|=k\\ i\in J}} \tilde{h}_{k-1}(x_{J\setminus\{i\}}) = 0.$$

(4) $$\sum_{c=0}^{m} \sum_{\substack{K\subset\{1,..,m\}\\ |K|=c}} h_c(x_K) =$$

$$= \sum_{c=0}^{m} \sum_{|K|=c} \sum_{k=0}^{c} (-1)^{c-k} \sum_{\substack{L\subset K\\ |L|=k}} \tilde{h}_k((x_K)_L) =$$

$$= \sum_{k=o}^{m} \sum_{\substack{L\subset\{1,\dots,m\}\\ |L|=k}} \tilde{h}_k(x_L) \sum_{c=k}^{m} (-1)^{c-k} \sum_{\substack{K\supset L\\ |K|=c}} 1.$$

Note that $\sum_{c=k}^{m} (-1)^{c-k} \sum_{\substack{K\supset L\\ |K|=c}} 1 = \sum_{i=o}^{m-k} \binom{m-k}{i}(-1)^i = \begin{cases} 1 & k=m \\ 0 & k<m, \end{cases}$

and hence

$$\sum_{c=o}^{m} \sum_{|K|=c} h_c(x_K) = \tilde{h}_m(x) = h(x)$$

by (1.2.2).

Theorem 1.2.1: (Decomposition Theorem)
Let h be a symmetric, unbiased estimator for ϑ of degree m. Then for any $Q \in M_o$ there exist kernels $h_c : E^c \to \mathbb{R}$ $(0 \leq c \leq m)$, defined in (1.2.3), satisfying the conditions of Lemma 1.2.1 such that for every $n \geq m$

$$(1.2.4)\quad U_n(h) = \sum_{c=o}^{m} \binom{m}{c} U_n(h_c).$$

Proof:
By Lemma 1.2.1, (4) we have that

$$U_n(h) = \binom{n}{m}^{-1} \sum_{1\leq i_1<\dots<i_m\leq n} \sum_{c=o}^{m} \sum_{1\leq k_1<\dots<k_c\leq m} h_c(X_{i_{k_1}},\dots,X_{i_{k_c}})$$
$$= \binom{n}{m}^{-1} \sum_{c=o}^{m} \sum_{1\leq i_1<\dots<i_m\leq n} \sum_{1\leq k_1<\dots<k_c\leq m} h_c(X_{i_{k_1}},\dots,X_{i_{k_c}}).$$

If $i_{k_1},\dots,i_{k_c}$ is fixed, then there exist exactly $\binom{n-c}{m-c}$ choices of different indices $i_{k_{c+1}},\dots,i_{k_m}$, also different from the fixed ones, in $\{1,\dots,n\}$. Hence interchanging the last two summations it follows that

$$U_n(h) = \binom{n}{m}^{-1} \sum_{c=o}^{m} \binom{n-c}{m-c} \sum_{1\leq i_1<\dots<i_c\leq n} h(X_{i_1},\dots,X_{i_c}),$$

and since $\binom{n}{m}^{-1}\binom{n-c}{m-c}\binom{n}{c} = \binom{m}{c}$, (1.2.4) follows.

Example 1.2.1:
The sample variance was defined in Example 1.1.2 using the

kernel $h(x,y) = \frac{1}{2}(x-y)^2$. From the definition (1.2.2) we therefore obtain (F denotes the corresponding d.f.)

$$\tilde{h}_1(x) = \int h(x,y)dF(y) = \frac{1}{2}x^2 - x\int ydF(y) + \\ + \frac{1}{2}\int y^2 dF(y)$$

$$h_1(x) = \frac{1}{2}(x - \int ydF(y))^2 - \frac{1}{2}\vartheta(F)$$

and $$h_2(x,y) = -(x - \int xdF(x))(y - \int ydF(y)).$$

Hence

$$U_n(h) = \vartheta(F) + \frac{1}{n}\sum_{i=1}^{n}[(X_i - \int ydF(y))^2 - \vartheta(F)] -$$

$$- \frac{2}{n(n-1)}\sum_{1\leq i<j\leq n}(X_i - \int ydF(y))(X_j - \int ydF(y)).$$

For the kernel $h(x,y) = 1_{\{x+y>0\}}$, used for the Wilcoxon statistic in Example 1.1.3 we obtain

$$\tilde{h}_1(x) = 1 - F(-x),\ h_1(x) = \tilde{h}_1(x) - \vartheta(F) \quad \text{and}$$

$$h_2(x,y) = h(x,y) - \tilde{h}_1(x) - \tilde{h}_1(y) + \vartheta(F).$$

In Example 1.1.4, $\tilde{h}_2(y,z) = \frac{1}{3}(1-F(y+z) + F(y-z) + F(z-y))$ and in Example 1.1.5

$$\tilde{h}_2(y,z) = 2/3\ (1/2 + F((y+z)/2) - F(2z-y) - F(2y-z)).$$

Lemma 1.2.2:

Let h be a degenerate kernel. Then $\{\binom{n}{m}U_n(h) : n \geq m\}$ is a martingale with respect to the filtration $\mathcal{A}_r = \sigma(X_1,\ldots,X_r)$ $(r \geq m)$.

Proof:

$$E(\binom{n}{m}U_n(h)|\mathcal{A}_{n-1}) = \sum_{1\leq i_1<\ldots<i_m\leq n-1} E(h(X_{i_1},\ldots,X_{i_m})|\mathcal{A}_{n-1}) +$$

$$+ \sum_{1\leq i_1<\ldots<i_{m-1}\leq n-1} E(h(X_{i_1},\ldots,X_{i_{m-1}},X_n)|\mathcal{A}_{n-1})$$

$$= \binom{n-1}{m} U_{n-1}(h) + \sum_{1 \le i_1 < \ldots < i_{m-1} \le n-1} \int h(X_{i_1},\ldots,X_{i_{m-1}},x)\,dQ(x)$$

$$= \binom{n-1}{m} U_{n-1}(h).$$

Lemma 1.2.3:

Let h and h' be two degenerate kernels (w.r. to $Q \in M_o$). If their degrees m and m' are different, then for every $n \geq m \vee m'$ the random variables $U_n(h)$ and $U_n(h')$ are orthogonal, provided the kernels are square integrable.

Proof:

Clearly,

$$E\, U_n(h)U_n(h') =$$

$$= \binom{n}{m}^{-1}\binom{n}{m'}^{-1} \sum_{1\le i_1<\ldots<i_m\le n} \; \sum_{1\le j_1<\ldots<j_{m'}\le n} E\, h(X_{i_1},\ldots$$

$$\ldots,X_{i_m})h'(X_{j_1},\ldots,X_{j_{m'}}) = 0$$

since every summand vanishes.

Lemma 1.2.4:

For a symmetric kernel h of degree m and for a probability $Q \in M_o$ assume that

$$(1.2.5)\quad \zeta_c := \mathrm{Var}\; \tilde{h}_c \qquad (1 \le c \le m)$$

and

$$(1.2.6)\quad \delta_c := E\, h_c^2 = \mathrm{Var}\; h_c \qquad (1 \le c \le m),$$

are finite, where $\tilde{h}_c$ and h_c are defined in (1.2.1) and (1.2.3). Then the relation between the ζ_c and δ_c is given by

$$(1.2.7)\quad \delta_c = \sum_{i=o}^{c-1} \binom{c}{i}(-1)^i \zeta_{c-i} \qquad (1\le c\le m)$$

and

$$(1.2.8)\quad \zeta_c = \sum_{i=o}^{c-1} \binom{c}{i}\delta_{c-i} \qquad (1\le c\le m).$$

Proof:

Using $\sum_{k=o}^{c} (-1)^{c-k} \binom{c}{k} = 0$ if $c \geq 1$ and the definition of h_c in (1.2.3) it follows that for $c \geq 1$

$$\delta_c = Eh_c^2(X^c) = E\Big(\sum_{k=o}^{c} (-1)^{c-k} \sum_K [\tilde{h}_k((X^c)_K) - \vartheta]\Big)^2$$

$$= \sum_{K,L\subset\{1,\ldots,c\}} (-1)^{|K|+|L|} E[(\tilde{h}_{|K|}((X^c)_K) - \vartheta)(\tilde{h}_{|L|}((X^c)_L) - \vartheta)]$$

$$= \sum_{K,L\subset\{1,\ldots,c\}} (-1)^{|K|+|L|} E[\tilde{h}_{|K\cap L|}((X^c)_{K\cap L}) - \vartheta]^2$$

$$= \sum_{j=1}^{c} \operatorname{Var} \tilde{h}_j \binom{c}{j} \sum_{k=o}^{c-j} \sum_{l=o}^{c-j-k} \binom{c-j}{k}\binom{c-j-k}{l}(-1)^{k+l}$$

$$= \sum_{j=1}^{c} \binom{c}{j}(-1)^{c-j} \zeta_j.$$

This is (1.2.7).

(1.2.8) will be shown by induction over c. For $c = 1$ $h_1 = \tilde{h}_1 - \vartheta$ by definition. Assume now that (1.2.8) holds for $c = 1,\ldots,r-1$ and we want to show it for $c = r$. By (1.2.7) and the hypothesis

$$\delta_r = \zeta_r + \sum_{i=1}^{r-1} \binom{r}{i}(-1)^i \zeta_{r-i}$$

$$= \zeta_r + \sum_{i=1}^{r-1} \binom{r}{i}(-1)^i \sum_{j=o}^{r-i-1} \binom{r-i}{j} \delta_{r-i-j}$$

$$= \zeta_r + \sum_{l=1}^{r-1} \delta_{r-l} \binom{r}{l} \sum_{i=1}^{l} (-1)^i \binom{l}{i}$$

$$= \zeta_r - \sum_{l=1}^{r-1} \binom{r}{l} \delta_{r-l}.$$

Examples of computing ζ_1 will be given in section 3.

Theorem 1.2.2:

Let h be a symmetric kernel of degree m and let $Q \in M_o$ satisfy $\int\ldots\int h^2(x_1,\ldots,x_m) \prod_{i=1}^{m} dQ(x_i) < \infty$.
If h is degenerate then

$$(1.2.9) \quad \operatorname{Var} U_n(h) = \binom{n}{m}^{-1} Eh^2(X_1,\ldots,X_m) = \binom{n}{m}^{-1} \delta_m \quad (n \geq m).$$

In general, we have

$$(1.2.10)\quad \operatorname{Var} U_n(h) = \binom{n}{m}^{-1} \sum_{c=1}^{m} \binom{m}{c}\binom{n-c}{m-c}\delta_c \qquad (n \geq m)$$

and

$$(1.2.11)\quad \operatorname{Var} U_n(h) = \binom{n}{m}^{-1} \sum_{c=1}^{m} \binom{m}{c}\binom{n-m}{m-c}\zeta_c \qquad (n \geq m),$$

where ζ_c and δ_c are defined in (1.2.5) and (1.2.6). (Note that we use (in (1.2.11)) the general definition for binomial coefficients $\binom{x}{q} = (q!)^{-1}x(x-1)\ldots(x-q+1)$.)

Proof:

First note that $\int\ldots\int h^2(x_1,\ldots,x_m) \prod_{i=1}^{m} dQ(x_i) < \infty$ implies that $\zeta_c < \infty$ for all $1 \leq c \leq m$ and hence also $\delta_c < \infty$ by Lemma 1.2.4. (1.2.9) follows immediately from

$$EU_n(h)^2 = \binom{n}{m}^{-2} \sum_{1\leq i_1<\ldots<i_m\leq n} \; \sum_{1\leq j_1<\ldots<j_m\leq n} Eh(X_{i_1},\ldots,X_{i_m})h(X_{j_1},\ldots,X_{j_m})$$

and

$$Eh(X_{i_1},\ldots,X_{i_m})h(X_{j_1},\ldots,X_{j_m}) = 0$$

if at least one i_k or j_l appears only once among all the indices $i_1,\ldots,i_m, j_1,\ldots,j_m$.

In order to show (1.2.10) it follows now from Lemma 1.2.3 and the Decomposition Theorem that

$$E(U_n(h)-\vartheta)^2 = E\Big(\sum_{c=1}^{m} \binom{m}{c} U_n(h_c)\Big)^2$$

$$= \binom{n}{m}^{-1} \sum_{c=1}^{m} \binom{m}{c}\binom{n-c}{m-c}\delta_c.$$

It is left to prove (1.2.11). From (1.2.8) we have

$$\zeta_c = \sum_{j=1}^{c} \binom{c}{j}\delta_j$$

and hence

$$\sum_{c=1}^{m} \binom{m}{c}\binom{n-m}{m-c}\zeta_c = \sum_{c=1}^{m} \delta_c \sum_{j=c}^{m} \binom{m}{j}\binom{j}{c}\binom{n-m}{m-j}$$

$$= \sum_{c=1}^{m} \binom{m}{c}\delta_c \sum_{k=o}^{m-c} \binom{m-c}{k}\binom{n-m}{m-c-k}.$$

Now apply the general formula (Feller (1950), p.48)

$$\sum_{k=o}^{b} \binom{a}{r-k}\binom{b}{k} = \binom{a+b}{r}$$

to see that the right-hand sides of (1.2.10) and (1.2.11) are equal.

Exercise: Prove (1.2.11) directly.

Theorem 1.2.3:

Let h be a symmetric kernel of degree m and let $Q \in M_o$. If $\zeta_m < \infty$ then

(1.2.12) $\quad 0 \leq d \operatorname{Var} \tilde{h}_c \leq c \operatorname{Var} \tilde{h}_d \qquad (1 \leq c \leq d \leq m)$

(1.2.13) $\quad \frac{m^2}{n} \operatorname{Var} \tilde{h}_1 \leq \operatorname{Var} U_n(h) \leq \frac{m}{n} \operatorname{Var} h \qquad (n \geq m)$

(1.2.14) $\quad \{n \operatorname{Var} U_n(h) : n \geq m\}$ is decreasing,

$\operatorname{Var} U_m(h) = \operatorname{Var} h$ and

$\lim_{n\to\infty} n \operatorname{Var} U_n(h) = m^2 \operatorname{Var} \tilde{h}_1.$

Proof:

(1.2.12): $c \operatorname{Var} \tilde{h}_d - d \operatorname{Var} \tilde{h}_c =$

$$= c \sum_{i=o}^{d-1} \binom{d}{i}\delta_{d-i} - d \sum_{j=o}^{c-1} \binom{c}{j}\delta_{c-j} =$$

$$= \sum_{i=1}^{c} [c\binom{d}{i} - d\binom{c}{i}]\delta_i + \sum_{j=c+1}^{d} c\binom{d}{j}\delta_j \geq 0$$

since $\frac{c}{d}\binom{d}{i}\binom{c}{i}^{-1} \geq 1$.

(1.2.13): Using (1.2.11) of Theorem 1.2.2, (1.2.12) in the form $c\zeta_1 \leq \zeta_c \leq c/m\ \zeta_m$ and the equality

$$(1.2.15) \quad \binom{n}{m}^{-1} \sum_{c=1}^{m} c\binom{m}{c}\binom{n-m}{m-c} = \frac{m^2}{n}$$

it follows immediately that

$$\operatorname{Var} U_n(h) \geq \binom{n}{m}^{-1} \sum_{c=1}^{m} c\binom{m}{c}\binom{n-m}{m-c}\zeta_1 = \frac{m^2}{n}\zeta_1$$

and

$$\operatorname{Var} U_n(h) \leq \binom{n}{m}^{-1} \sum_{c=1}^{m} \frac{c}{m}\binom{m}{c}\binom{n-m}{m-c}\zeta_m = \frac{m}{n}\zeta_m .$$

(1.2.14): Again using (1.2.11) we have

$$n \operatorname{Var} U_n(h) - (n+1) \operatorname{Var} U_{n+1}(h) =$$

$$= \sum_{c=1}^{m} \zeta_c [n\binom{n}{m}^{-1}\binom{m}{c}\binom{n-m}{m-c} - (n+1)\binom{n+1}{m}^{-1}\binom{m}{c}\binom{n+1-m}{m-c}]$$

$$= \sum_{c=1}^{m} \zeta_c \binom{m}{c}\binom{n+1-m}{m-c}\binom{n}{m}^{-1} \frac{1}{n+1-m} [n(c-1) - (m-1)^2].$$

Choose c_o satisfying $n(c-1) - (m-1)^2 \leq 0$ for $c \leq c_o$ and > 0 for $c > c_o$. By (1.2.12)

$$- \zeta_c \geq - \frac{c}{c_o}\zeta_{c_o} \qquad (c \leq c_o)$$

$$\zeta_c \geq \frac{c}{c_o}\zeta_{c_o} \qquad (c > c_o)$$

and therefore

$$n \operatorname{Var} U_n(h) - (n+1)\operatorname{Var} U_{n+1}(h) \geq$$

$$\geq c_o^{-1} \zeta_{c_o} \sum_{c=1}^{m} \binom{m}{c}\binom{n+1-m}{m-c}\binom{n}{m}^{-1} \frac{c}{n+1-m} (n(c-1) - (m-1)^2)$$

$$= c_o^{-1} \zeta_{c_o} \{\sum_{c=1}^{m} nc\binom{n}{m}^{-1}\binom{m}{c}\binom{n-m}{m-c} - \sum_{c=1}^{m} (n+1)c\binom{n+1}{m}^{-1}\binom{m}{c}\binom{n+1-m}{m-c}\}$$

$= 0$ by (1.2.15)

If $\zeta_1 > 0$ the variance of $U_n(h)$ is of order n, the same magnitude as for the variance of the partial sums of i.i.d. random variables. If $\zeta_1 = 0$ this is no longer true. While in the first case we shall prove the asymptotic normality for U-statistics in the next section, the second case will give other types of limit laws, discussed in chapter 2 .

Let us note at the end of this section the following

Theorem 1.2.4:

Suppose h is a symmetric kernel of degree m and $Q \in M_o$. If $h_c \equiv 0$ $(1 \leq c \leq d-1)$ for some $1 \leq d \leq m$, where h_c is defined in (1.2.3), then for some constant K and for all $n \geq m$

$$(1.2.16) \quad E(\tbinom{m}{d}U_n(h_d) - (U_n(h) - \vartheta))^2 \leq K\, n^{-d-1},$$

provided $\xi_m < \infty$.

Proof:

Since $h_c \equiv 0$ for $c < d$ one obtains from the Decomposition Theorem and Lemma 1.2.3 that

$$E(\tbinom{m}{c}U_n(h_d) - (U_n(h)-\vartheta))^2 = \sum_{c=d+1}^{m} \tbinom{m}{c}^2 \tbinom{n}{c}^{-1} \delta_c$$

$$\leq n^{-d-1} (2m)^m \sum_{c=d+1}^{m} \tbinom{m}{c}^2 \delta_c.$$

Note that the constant K can be estimated in terms of ζ_m, since by (1.2.7), (1.2.8) and (1.2.12)

$$\delta_c \leq \sum_{i=d}^{c} \tbinom{c}{i} \frac{i}{m} \zeta_m.$$

Remark:

The condition $h_c \equiv 0$ for $1 \leq c < d$ is equivalent to $\delta_c = 0$ for $1 \leq c < d$ and this is by (1.2.8) equivalent to $\zeta_c = 0$ $(1 \leq c < d)$.

3. Convergence theorems for U-statistics

This section contains a detailed discussion of weak convergence results for non-degenerate U-statistics. The standing assumptions are kept as in the last two sections: h denotes a kernel of degree m and $P \in M_o$. In addition we shall assume that $\int \ldots \int h^2(x_1,\ldots,x_m) \prod_{i=1}^{m} dP(x_i) < \infty$, equivalently $\zeta_m < \infty$, except otherwise stated.

We first note that Theorem 1.2.4 immediately implies the following

Proposition 1.3.1:

Suppose the kernel h satisfies $h_c \equiv 0$ $(1 \leq c < d)$ for some $1 \leq d \leq m$. Then $n^{d/2} \binom{m}{d} U_n(h_d)$ converges weakly to a measure μ if and only if $n^{d/2}(U_n(h)-\vartheta)$ converges weakly to μ.

If $d = 1$ then $h_1(x) = \tilde{h}_1(x) - \vartheta$ and hence

$$\sqrt{n}\; m\, U_n(h_1) = \frac{m}{\sqrt{n}} \sum_{i=1}^{n} (\tilde{h}_1(X_i) - \vartheta)$$

is a sum of i.i.d. random variables normalized by $m^{-1} n^{1/2}$. Since $E\,\tilde{h}_1^2(X_1) = \zeta_1 + \vartheta^2 < \infty$ the central limit theorem applies. Thus,

Theorem 1.3.1:

For any kernel h of degree m with $\zeta_m < \infty$ $\sqrt{n}\,(U_n(h) - \vartheta)$ converges weakly to $N(0, m^2\zeta_1)$.

Remark:

Here and in the following $N(\mu,\sigma^2)$ denotes the normal distribution on $\mathbb{R}$ with expectation μ and variance σ^2. If $\sigma^2 = 0$, then weak convergence to $N(\mu,0)$ is just convergence in probability to μ. Thus, $\sqrt{n}\,(U_n(h) - \vartheta)$ converges weakly to zero if $\zeta_1 = 0$, equivalently $h_1 \equiv 0$ or $\tilde{h}_1 \equiv \vartheta$. We note that Theorem 1.2.4 also implies that $\lim_{n\to\infty} n\, E(U_n(h) - \vartheta)^2 = m^2\, \zeta_1$. This additional property, which is not a consequence of weak convergence, will be discussed in more detail in chapters 3 and 4.

Example 1.3.1:

Let us first compute the asymptotic variances of some of the examples in the first section. For the k-th moment asymptotic normality holds if $E|X_1|^{2k} < \infty$ and the asymptotic variance is given by $\operatorname{Var} X_1^k$. Similarly, the sample variance is asymptotically normal $N(0,4\,\zeta_1)$, provided $E\,X_1^4 < \infty$. Here $h_1(x) = 1/2\,(x - \int y\, dP(y))^2 - 1/2 \int (z - \int y\, dP(y))^2 dP(z)$

and hence $4\,\zeta_1 = E(X_1-EX_1)^4 - (\mathrm{Var}\,X_1)^2$. If, for example, $X_1 \sim N(\mu,\sigma^2)$ then $4\,\zeta_1 = 2\sigma^4$.

The Wilcoxon signed rank statistic W^+ (cf. Example 1.1.3) can be written as $W^+ = n\,U_n(h_1) + \binom{n}{2}U_n(h_2)$, where $h_1(x) = 1_{(0,\infty)}(x)$ and $h_2(x,y) = h_1(x+y)$. By Theorem 1.2.2

$$\mathrm{Var}(\binom{n}{2}^{-1} n^{3/2} U_n(h_1)) = \binom{n}{2}^{-2} n^2\ \mathrm{Var}\,h_1 \to 0,$$

hence we deduce from Theorem 1.3.1 that

$$\sqrt{n}\ \binom{n}{2}^{-1} (W^+ - EW^+) \to N(0,\sigma^2)$$

weakly, where $\sigma^2 = 4\,\mathrm{Var}(\widetilde{h_2})_1$. Note that according to (1.2.1) $(\widetilde{h_2})_1 = \int h_2(x,y)dP(y) = 1 - F_\vartheta(-x)$. Under H_0 (i.e., $\vartheta = 0$ and $F = F_0$ symmetric) $\sigma^2 = 4(\int_0^1 y^2dy - (\int_0^1 y\,dy)^2) = 1/3$, showing that the Wilcoxon
signed rank statistic under H_0 is asymptotically distribution-free. Similarly, the Wilcoxon one sample-test is asymptotically distribution-free under H_0.

Let us consider the U-statistic of Example 1.1.4 under H_0 (testing for exponential distribution). Clearly, under H_0, the distribution of $U_n(h)$ is independent of the parameter of the exponential distribution. The asymptotic variance is given by $9\,\zeta_1$ where $\tilde{h}_1(x) = 1/3\,(1-x\,e^{-x})$. It follows that $\zeta_1 = \int \tilde{h}_1(x)^2\, e^{-x}\,dx - 1/16 = 5/3888$ and $\sigma^2 = 9\,\zeta_1 = 5/432$.

Let $X_1,X_2,\ldots$ be i.i.d. $\mathbb{R}^2$-valued vectors, $X_i = (X_{i,1},X_{i,2})$, with continuous d.f. $F : \mathbb{R}^2 \to [0,1]$.
X_i and X_j $(i\neq j)$ are called concordant, if $(X_{j,2}-X_{j,1})(X_{i,2}-X_{i,1}) > 0$. The number of concordant pairs among $X_1,\ldots,X_n$ obviously defines a U-statistic $U_n(h)$ with kernel $h : (\mathbb{R}^2)^2 \to \mathbb{R}$ defined by

$h((x_1,x_2),(y_1,y_2)) = 1_{\{(x_2-y_2)(x_1-y_1)>0\}}$. $2\,U_n(h) - 1$ is known as *Kendall's sample correlation coefficient*. $U_n(h)$ estimates the parameter $\vartheta = P((X_{2,2}-X_{1,2})(X_{2,1}-X_{1,1}) > 0)$. If $X_{i,1}$ and $X_{i,2}$ are independent, equivalently if $F(x,y) = F_1(x)F_2(y)$, then

$$\vartheta = 2 \int 1_{\{x>y\}} dF_2(x)dF_2(y) \int 1_{\{x>y\}} dF_1(x)dF_1(y) = 1/2.$$

$U_n(h)$ is used to test the hypothesis H_o of independence of $X_{i,1}$ and $X_{i,2}$ against dependence ($\vartheta \neq 1/2$). Clearly, $\sqrt{n}\,(U_n(h) - \vartheta)$ is asymptotically normal with variance $\sigma^2 = 4\,\zeta_1$. Under H_o, $\tilde{h}_1(x_1,x_2) =$

$$\int\int 1_{\{x_2>y_2,x_1>y_1\}} dF_1(y_1)dF_2(y_2) +$$

$$+ \int\int 1_{\{x_2<y_2,x_1<y_1\}} dF_1(y_1)dF_2(y_2) = F_1(x_1-)F_2(x_2-) +$$

$$+ (1 - F_1(x_1))(1-F_2(x_2)).$$

If F_1 and F_2 are continuous, then

$$\zeta_1 = \int_o^1 \int_o^1 (xy+(1-x)(1-y))^2 dx\, dy - 1/4 =$$

$$= 1/3 \int_o^1 (xy + (1-x)(1-y))^3 \Big|_o^1 dy - 1/4 = 1/12,$$

hence $\sigma^2 = 1/3$.

Example 1.3.2:

We shall briefly discuss asymptotically distribution-free confidence intervals for U-statistics, at first the case of a known asymptotic variance and secondly the case when the asymptotic variance has to be estimated consistently.

Consider the Wilcoxon signed rank statistic W^+. According to the previous example, under H_o,

$$\frac{W^+ - \frac{n(n+1)}{4}}{\sqrt{\frac{n(n-1)^2}{12}}} \to N(0,1).$$

Fix $\alpha > 0$ and define $d_1 = \frac{n(n+1)}{4} - \Phi^{-1}(1 - \alpha/2)\sqrt{\frac{n(n-1)^2}{12}}$ and $d_2 = \frac{n(n+1)}{4} + \Phi^{-1}(1 - \alpha/2)\sqrt{\frac{n(n-1)^2}{12}}$, i.e.

$$\lim_{n\to\infty} P(d_1 \leq W^+ \leq d_2) = 1 - \alpha .$$

We want to find functions l and u such that for every

$\vartheta \in \mathbb{R}$ and every $x = (x_1,\dots,x_n)$ $d_1 \leq W^+((x_1-\vartheta,\dots,x_n-\vartheta)) \leq$ $\leq d_2 \Leftrightarrow l(x_1,\dots,x_n) \leq \vartheta < u(x_1,\dots,x_n)$. Then $[l(x),u(x))$ is the approximate $1-\alpha$ confidence interval for ϑ, i.e. for all ϑ

$$\lim_{n\to\infty} P_\vartheta^n(\{x : l(x) \leq \vartheta < u(x)\}) = 1 - \alpha.$$

We have

$$d_1 \leq W^+((x_1-\vartheta,\dots,x_n-\vartheta)) \leq d_2 \Leftrightarrow$$

$$\Leftrightarrow |\{i \leq j : x_i + x_j > 2\vartheta\}| \in [d_1,d_2].$$

Let $L_{(1)} \leq L_{(2)} \leq \dots \leq L_{(\frac{n(n+1)}{2})}$ denote the ordered values of $x_i + x_j$ $(1 \leq i \leq j \leq n)$ and let k denote the index defined by $k = \min\{j : L_{(j)} > 2\vartheta\}$. Then $|\{i \leq j : x_i + x_j > 2\vartheta\}| =$ $\frac{n(n+1)}{2} - k + 1$, $L_{(k-1)} \leq 2\vartheta < L_{(k)}$ and $\frac{n(n+1)}{2} - k + 1 \in [d_1,d_2]$ iff $\frac{n(n+1)}{2} + 1 - d_2 \leq k \leq$ $\leq \frac{n(n+1)}{2} + 1 - d_1$. It follows that $d_1 \leq W^+ \leq d_2$ iff $L_{(\frac{n(n+1)}{2} - d_2)} \leq 2\vartheta < L_{(\frac{n(n+1)}{2}+1-d_1)}$, determining the bounds for the confidence interval explicitly.

For the case of an unknown asymptotic variance consider the simple example of the sample mean $T_n = \sum_{i=1}^{n} X_i$. Under H_o (i.e. $\vartheta = 0$ in the location model $\{F_\vartheta : F_\vartheta(x) = F(x-\vartheta)\}$ $\frac{1}{\sqrt{n}} T_n$ converges weakly to $N(0,\sigma^2)$ where σ^2 is unknown. Let $\hat{\sigma}^2$ be a consistent estimator for σ^2. Then $\frac{1}{\sqrt{n}\,\hat{\sigma}} T_n \to N(0,1)$, so that - given $\alpha > 0$ -

$$d_1 = -\Phi^{-1}(1-\tfrac{\alpha}{2})\ \sqrt{n}\ \hat{\sigma} \quad \text{and} \quad d_2 = \Phi^{-1}(1-\tfrac{\alpha}{2})\ \sqrt{n}\ \hat{\sigma}\ .$$

Clearly in this case the confidence interval is given by $[\sum X_i - d_2, \sum X_i - d_1]$.

Example 1.3.3:

What has been said in the last example concerning the construction of asymptotically distribution-free confidence intervals, basically also applies to test problems. If the asymptotic variance of a statistic does not depend (under H_o) on the distribution of the X_i's, then for given level α the critical region of a level-α-test based on the statistic $U_n(h)$ is determined by the normal distribution. Otherwise the asymptotic variance should be estimated consistently to obtain an asymptotic distribution-free statistic. This is carried out very much in the same way as in Example 1.3.2. For example, testing the parameter ϑ in the location model $\{F_\vartheta : \vartheta \in \mathbb{R}\}$ for $H_o = \{\vartheta \leq 0\}$ against $\vartheta > 0$, the critical region, when using the Wilcoxon signed rank test, will be given by $\{W^+ > K\}$ where K satisfies

$$K = \frac{n(n+1)}{4} - \Phi^{-1}(1-\alpha)\sqrt{\frac{n(n-1)^2}{12}} .$$

Using the martingale property of a U-statistic in Lemma 1.1.1 or in Lemma 1.2.2, Theorem 1.3.1 easily extends to an invariance principle for non-degenerate U-statistics. All that is needed is a maximal inequality for martingales. Here we use Chow's inequality.

Lemma 2.3.1:

Let $S_1, S_2, \ldots, S_N$ be a submartingale with respect to the filtration $\mathcal{F}_1, \ldots, \mathcal{F}_N$, i.e. $S_n \leq E(S_{n+1} | \mathcal{F}_n)$. If $a_1 \geq a_2 \geq \ldots$ $\ldots \geq a_N > 0$, then for every $\varepsilon > 0$

$$(1.3.1)\quad \varepsilon P(\max_{1\leq n\leq N} a_n S_n \geq \varepsilon) \leq a_1 ES_1^+ + \sum_{n=2}^{N} a_n E(S_n^+ - S_{n-1}^+) - a_N \int_{\{\max a_n S_n < \varepsilon\}} S_N^+ \, dP$$

and

$$(1.3.2)\quad \varepsilon P(\max_{1\leq n\leq N} a_n S_n \geq \varepsilon) \leq \sum_{n=1}^{N-1} (a_n - a_{n+1}) ES_n^+ + a_N ES_N^+ .$$

Proof:

(1.3.2) follows easily from (1.3.1), hence it suffices to show the first inequality.

Setting $A = \{\max a_n S_n \geq \varepsilon\}$ and $A_n = \{a_i S_i < \varepsilon, a_n S_n \geq \varepsilon, i<n\}$ and using $a_n S_n \geq \varepsilon > 0$ on A_n one obtains

$$\varepsilon P(A) = \varepsilon \sum_{n=1}^{N} P(A_n) \leq \sum_{n=1}^{N} a_n \int_{A_n} S_n^+ dP$$

$$= a_1 ES_1^+ - a_1 \int_{A_1^c} S_1^+ dP + \sum_{n=2}^{N} a_n \int_{A_n} S_n^+ dP.$$

Because of

$$- a_1 \int_{A_1^c} S_1^+ dP + a_2 \int_{A_2} S_2^+ dP \leq$$

$$\leq a_2 \int_{A_1^c} (S_2^+ - S_1^+) dP - a_2 \int_{(A_2 \cup A_1)^c} S_2^+ dP,$$

it follows that

$$\varepsilon P(A) \leq a_1 ES_1^+ + a_2 \int_{A_1^c} (S_2^+ - S_1^+) dP -$$

$$- a_2 \int_{(A_1 \cup A_2)^c} S_2^+ dP + \sum_{n=3}^{N} a_n \int_{A_n} S_n^+ dP.$$

Iteration of this procedure yields

$$\varepsilon P(A) \leq a_1 ES_1^+ + \sum_{n=2}^{N} a_n \int_{(A_1 \cup .. \cup A_{n-1})^c} (S_n^+ - S_{n-1}^+) dP$$

$$- a_N \int_{A^c} S_N^+ dP.$$

(1.3.1) follows now from the submartingale property.

The random functions for which we shall prove the weak invariance principle (the extension of Theorem 1.3.1) are defined by

$$(1.3.3)\quad Y_n(t) = \begin{cases} (n\, m^2\, \zeta_1)^{-1/2} k(U_k(h) - \vartheta) & \text{if } t = \frac{k}{n},\ m \leq k \leq n \\ 0 & t \leq \frac{m-1}{n} \\ \text{by linear interpolation elsewhere,} & \end{cases}$$

if $\zeta_1 > 0$. Clearly, Y_n is a random function with values in $C([0,1])$, the space of continuous functions on $[0,1]$.

Recall that on $C([0,1])$ the Borel-σ-algebra is given by the topology of uniform convergence, which is metrizable by $\|f\|_\infty = \sup_{0\le t\le 1} |f(t)|$. We denote by $D([0,1])$ the space of all right continuous functions on $[0,1]$ for which the left limits exist. Here the topology is given by the Skorohod-metric, which is weaker than the supremum's-metric. The random functions in $D([0,1])$ we are concerned with are defined by

$$(1.3.4)\quad Z_n(t) = Y_n\left(\frac{[tn]}{n}\right),$$

where $[\alpha]$ denotes the integer part of α. Finally, we denote by $\{W(t) : 0 \le t \le 1\}$ the standard Wiener process on $C([0,1])$ (also regarded as a random element in $D([0,1])$).

Theorem 1.3.2:

Let h be a kernel of degree m, $\zeta_m < \infty$ and $\zeta_1 > 0$. Then $\{Y_n : n \ge m\}$, (resp. $\{Z_n : n \ge m\}$) converges weakly in $C([0,1])$, (resp. $D([0,1])$) to the Wiener process W.

Proof:

By Donsker's theorems (Billingsley 1968, p.68 and p.137)

$$\tilde{Y}_n(t) = \begin{cases} (n\,\zeta_1)^{-1/2} \sum_{i=1}^{k} (\tilde{h}_1(X_i)-\vartheta) & \text{if } t = \frac{k}{n}\ (0\le k\le n) \\ \text{by linear interpolation elsewhere.} \end{cases}$$

(resp. $\tilde{Z}_n(t) = \tilde{Y}_n(\frac{[tn]}{n})$) $(n\ge 1)$ converge weakly in $C([0,1])$, (resp. $D([0,1])$) to W. Hence it suffices to show that $Y_n-\tilde{Y}_n$ (resp. $Z_n-\tilde{Z}_n$) converge to zero in probability with respect to the supremum metric $\|\ \|_\infty$. Since it turns out that for $Z_n-\tilde{Z}_n$ this amounts to the same estimate as for $Y_n-\tilde{Y}_n$, it is sufficient to prove

$$\lim_{n\to\infty} P\left(\sup_{k\le n}\left|Y_n\left(\frac{k}{n}\right) - \tilde{Y}_n\left(\frac{k}{n}\right)\right| \ge \varepsilon\right) = 0$$

for every $\varepsilon > 0$. By the Decomposition Theorem 1.2.1 we can reduce this again to estimate

$$P\Big(\sup_{0\leq k<m} (n\zeta_1)^{-1/2} \Big| \sum_{i=1}^{k} (\tilde{h}_1(X_i)-\vartheta) \Big| \geq \varepsilon \Big)$$

and

$$P\Big(\sup_{m\leq k\leq n} (nm^2\zeta_1)^{-1/2} k \Big| \sum_{c=2}^{m} \binom{m}{c} U_k(h_c) \Big| \geq \varepsilon \Big).$$

The first term clearly is bounded by $\varepsilon^{-2}m^2n^{-1}$ and the second one by the sum of terms

$$(1.3.5) \quad P\Big(\sup_{m\leq k\leq n} (nm^2\zeta_1)^{-1/2} k \binom{m}{c} \big|U_k(h_c)\big| \geq \frac{\varepsilon}{m-1} \Big)$$

$$(c=2,\dots,m).$$

Since $\binom{k}{c} U_k(h_c)$ is a martingale, $\binom{k}{c}^2 U_k(h_c)^2$ is a sub-martingale. Now apply Chow's inequality with $a_k = k^2 \binom{k}{c}^{-2}$ (indeed $a_k - a_{k+1} = \binom{k}{c}^{-2} [k^2-(k+1-c)^2] > 0$). It follows that (1.3.5) is bounded by

$$\varepsilon^{-2} \binom{m}{c}^2 \Big(\frac{m-1}{m}\Big)^2 n^{-1} \zeta_1^{-1} \delta_c \Big[\sum_{k=m}^{n-1} \binom{k}{c}^{-1} (2k(c-1)-(c-1)^2) + n^2 \binom{n}{c}^{-1} \Big]$$

where Theorem 1.2.2 (1.2.9) is used together with (1.3.2). Clearly, this tends to zero as $n \to \infty$ and the claim is proved.

<u>Example 1.3.4:</u>
Let h be a kernel of degree m and assume that $\zeta_m < \infty$ and $\zeta_1 > 0$. Assume the experimenter chooses the sample size n at random, according to random variables N_n satisfying $n^{-1}N_n \to Z$ in probability (for example, if the data are incomplete), where Z is a positive, bounded random variable. Let us define

$$\tilde{Z}_n(t) = \begin{cases} (N_n m^2 \zeta_1)^{-1/2} [N_n t] (U_{[N_n t]}(h) - \vartheta) & \text{if } t \geq m\, N_n^{-1} \\ 0 & \text{otherwise} \end{cases}$$

We will show below that $\tilde{Z}_n$ converges weakly in $D([0,1])$ to the Wiener process W. As a special case, for $t=1$, it follows that $N_n^{1/2}(U_{N_n}(h)-\vartheta) \longrightarrow N(0,m^2\zeta_1)$ weakly.

For the proof of the weak convergence we can follow Billingsley (1968, p.147).

Changing N_n by multiplication, we may assume that $0 < Z \leq K < 1$. Define

$$\psi_n(t) = \begin{cases} t\,n^{-1}\,N_n & \text{if} \quad N_n \leq n \\ tZ & \text{if} \quad N_n > n, \end{cases}$$

so that by assumption ψ_n converges in probability in $D([0,1])$ to the random element ψ, given by $\psi(t) = tZ$. Also the variables $(\psi_n, n^{-1}N_n)$ converge in probability (w.r. to the product topology of $D([0,1]) \times \mathbb{R}$) to (ψ, Z). Recall now Theorem 4.5 in Billingsley (1968, p.27): "Suppose that X' and X'' are independent and that X'' has the same distribution as Y''. If $X_n'' \to Y''$ in probability, and if $P(\{X_n' \in A'\} \cap E) \to P(\{X' \in A'\})P(E)$ for each X'-continuity set A' and each measurable set $E \in \sigma(X_n' : n \geq 1)$, then $(X_n', X_n'') \to (X', X'')$ weakly." We will apply this theorem with $X_n'' = (\psi_n, n^{-1}N_n), Y'' = (\psi, Z), X_n' = Z_n$ as defined in (1.3.4), $X' = W$, the Wiener process, and with $X'' = (\psi_o, Z_o)$ independent of W, where (ψ_o, Z_o) and (ψ, Z) have the same distribution. Hence we have to verify the "independence" condition, and this certainly only for cylinder sets $D = \{(X_1, \ldots, X_k) \in C\}$ where $C \subset E^k$ and $k \geq 1$. Fix $k \geq 1$. Replace in the definition of Z_n the sequence $X_1, X_2, X_3, \ldots$ by the sequence $X_{k+1}, X_{k+2}, \ldots$ and denote this random element by $\bar{Z}_n$. Then $\bar{Z}_n$ and 1_D are independent and $Z_n - \bar{Z}_n$ converges to zero in probability w.r. to $D([0,1])$. (This last fact should be verified as an exercise!) Hence, $\bar{Z}_n \to W$ weakly and $P(\{Z_n \in A'\} \cap D) - P(\{\bar{Z}_n \in A'\} \cap D) \to 0$. This proves that $(Z_n, \psi_n, n^{-1}N_n)$ converges weakly to (W, ψ_o, Z_o) w.r. to $C([0,1])^2 \times \mathbb{R}_+$. The map $(f, \varphi, \alpha) \to \alpha^{-1/2} f \circ \varphi$ defined on $D([0,1]) \times C([0,1]) \times \mathbb{R}_+$ with values in $D([0,1])$ is continuous for every point $(f, \varphi, \alpha) \in C([0,1])^2 \times \mathbb{R}_+$. Hence, by Corollary 1 to Theorem 5.1, $(n^{-1}N_n)^{-1/2} Z_n \circ \psi_n \to Z_o^{-1/2} W \circ \psi_o$ weakly. Observe that the distribution of

$Z_o^{-1/2}\, W(t\, Z_o)$ is the same as that of W and that

$$(n^{-1}N_n)^{-1/2}\, Z_n o\psi_n(t) = (N_n\, m^2\, \zeta_1)^{-1/2}[n\, \psi_n(t)](U_{[n\psi_n(t)]}(h)-\vartheta)$$

$$= \tilde{Z}_n(t) \quad \text{if} \quad N_n \leq n.$$

Since $K < 1$, $n^{-1}N_n \leq 1$ holds with a probability which goes to 1.

The assumption that Z is bounded is not essential. Knowing the bounded case, a general Z can be treated as in Billingsley (1968, p.148).

The essential point in the proof of Theorem 1.3.2 is the fact that for $c \geq 2$, $k^{1/2}U_k(h_c)$ converges to zero in probability. In fact, we can make a much stronger statement.

Theorem 1.3.3:
Let h be a degenerate kernel of degree m and $\zeta_m < \infty$. Then for any $\varepsilon > 0$

$$\lim_{n\to\infty} n^{\frac{m}{2}-\varepsilon}\; U_n(h) = 0 \quad \text{a.s.} \tag{1.3.6}$$

Proof: W.l.o.g. $\varepsilon < 1/2$.
Given $\eta > 0$ it is sufficient to show that

$$\lim P(A_n) = 0$$

where $A_n = \{\sup_{k\geq n} k^{\frac{m}{2}-\varepsilon}\, |U_k(h)| \geq \eta\}$. Fix $m \leq n < N$ and set

$$A_{n,N} = \{\sup_{n\leq k\leq N} k^{\frac{m}{2}-\varepsilon}\, |U_k(h)| \geq \eta\}.$$

Since $U_N(h),\ldots,U_n(h)$ is a martingale by Lemma 1.1.1, its squares form a submartingale. Moreover, setting $a_k = k^{m-2\varepsilon}$ we have $a_{k+1}-a_k > 0$ and hence by Chow's inequality again

$$P(A_{n,N}) \leq \eta^{-2}\Big(\sum_{k=n+1}^{N} (a_{k+1}-a_k)EU_{k+1}(h)^2 + a_n EU_n(h)^2\Big)$$

$$\eta^{-2}\zeta_m[n^{m-2\varepsilon}\binom{n}{m}^{-1} + \sum_{k=n+1}^{\infty} (m-2\varepsilon-1)(k+1)^{m-2\varepsilon-1}\binom{k+1}{m}^{-1}].$$

This bound is independent of N and since $A_{n,N} \nearrow A_n$, $P(A_n) = O(n^{-2\varepsilon})$, proving the theorem.

As a consequence of the preceding theorem we also have an almost sure invariance principle.

Theorem 1.3.4:

Let h be a kernel of degree m and $\zeta_m < \infty$. Then there exists an extension of the probability space such that there exists a standard Brownian motion $\{B(t) : t \geq 0\}$ on this richer space satisfying

$$(1.3.7) \quad \lim_{t\to\infty} (t \log\log t)^{-\frac{1}{2}} \left| \zeta^{-\frac{1}{2}} m^{-1} t(U_{[t]}(h)-\vartheta) - B(t) \right| = 0 \text{ a.s.}$$

Proof:

By Theorem 1.3.3 and the Decomposition Theorem

$$\sqrt{\frac{n}{\zeta_1 m^2 \log\log n}}(U_n(h)-\vartheta) - \frac{1}{\sqrt{n\zeta_1 \log\log n}} \sum_{i=1}^{n} (\tilde{h}_1(X_i)-\vartheta)$$

$$= \sum_{c=2}^{m} \binom{m}{c} \zeta_1^{-1/2} m^{-1} \sqrt{\frac{n}{\log\log n}} U_n(h_c) \to 0 \quad \text{a.s.} .$$

The classical almost sure invariance principle (cf. Gaenssler, Stute (1977), p. 397) ensures the extension of the probability space (for example, by an independent copy of $([0,1],\lambda)$, where λ denotes the Lebesgue measure) such that on the larger probability space one can define a Brownian motion $B(t)$ $(t \geq 0)$ satisfying

$$\frac{1}{\sqrt{t \log\log t}} \left| \sum_{i=1}^{[t]} \zeta_1^{-1/2}(\tilde{h}_1(X_i)-\vartheta) - B(t) \right| \longrightarrow 0 \quad \text{a.s.}$$

The theorem follows.

An immediate consequence is the law of the iterated logarithm:

Theorem 1.3.5:

Let h be a kernel of degree m with $\zeta_m < \infty$. Then

$$\limsup_{n\to\infty} (n \log\log n)^{-1/2} n(U_n(h)-\vartheta) = m\, \zeta_1^{1/2} \text{ a.s..}$$

All the theorems so far transfer to V-statistics, defined in Definition 1.1.3. Most of them follow from this:

Proposition 1.3.2:

Let h be a kernel of degree m satisfying

$$(1.3.8)\quad \max\{E\left|h(X_{i_1},\ldots,X_{i_m})\right| : 1\le i_1,\ldots,i_m\le m\} < \infty.$$

Then

$$(1.3.9)\quad \lim_{n\to\infty} \sqrt{n}(U_n(h)-V_n(h)) = 0 \quad \text{a.s. and in } L_1 .$$

Moreover,

$$(1.3.10)\quad \max_{m\le k\le n} n^{-1/2}(kU_k(h)-k\,V_k(h))\longrightarrow 0$$

a.s. as $n\longrightarrow\infty$.

Proof:

First note that

$$(1.3.11)\quad U_n(h) - V_n(h) = U_n(h)\left(1-\frac{m!}{n^m}\binom{n}{m}\right) - n^{-m}\sum h(X_{i_1},\ldots,X_{i_m})$$

where the summation extends over all choices of indices $1\le i_1,\ldots,i_m\le n$ for which at least two of them are equal. Since $1 - m!\,n^{-m}\binom{n}{m} = O(n^{-1})$, it follows from Theorem 1.1.2 that $\sqrt{n}\,U_n(h)(1-m!\,n^{-m}\binom{n}{m})\longrightarrow 0$ a.s. and in L_1.

The second summand on the right-hand side in (1.3.11) can be written as a sum of U-statistics with kernels in less than m arguments as follows:

For a partition $I = \{m_1,m_1+m_2,\ldots,m=m_1+\ldots+m_c\}$ of $\{1,\ldots,m\}$, where $m_i\ge 2$ for some $1\le i\le c$, define

$$g_I(x_1,\ldots,x_c) = h(x_1,\ldots,x_1,x_2,\ldots,x_c)$$

where x_i $(1\le i\le c)$ appears m_i times successively as an argument in h.

Then the second sum equals

$$n^{-m} \sum_{I} \frac{m!}{m_1!\dots m_c!} \binom{n}{c} U_n(g_I).$$

Each g_I is symmetric. To see this fix indices i and j. Then

$$g_I(x_1,\dots,x_c) = h(x_1,\dots,x_{i-1},\underbrace{x_i,\dots,x_i}_{m_i},x_{i+1},\dots$$
$$x_{j-1},\underbrace{x_j,\dots,x_j}_{m_j},\dots,x_c)$$

and

$$g_I(x_1,\dots,x_{i-1},x_j,x_{i+1},\dots,x_{j-1},x_i,x_{j+1},\dots,x_c) =$$
$$= h(x_1,\dots,x_{i-1},\underbrace{x_j,\dots,x_j}_{m_j},x_{i+1},\dots,x_{j-1},\underbrace{x_i,\dots,x_i}_{m_i},x_{j+1},\dots x_c).$$

The arguments in both right-hand sides are a permutation of each other, and since h is symmetric the right-hand sides coincide. Thus it is left to show that

$$n^{-m+1/2} \binom{n}{c} U_n(g_I) \longrightarrow 0 \quad \text{a.s. and in } L_1. \tag{1.3.11}$$

It follows from Theorem 1.1.2 that $U_n(g_I) \longrightarrow Eh(X_1,\dots,X_1,X_2,\dots,X_c)$ a.s. and in L_1. Since $c < m$,

$$n^{-m+1/2} \binom{n}{c} \leq n^{-m+c+1/2} \leq n^{-1/2}.$$

Therefore (1.3.11) holds for each I which is not a partition into points. This proves (1.3.9).

(1.3.10) follows immediately from (1.3.9).

The following results are immediate corollaries to the last proposition.

<u>Theorem 1.3.6:</u> Let h be a kernel ot degree m satisfying (1.3.8). Then

$$V_n(h) \longrightarrow \vartheta$$

a.s. and in L_1 as $n \longrightarrow \infty$.

<u>Proof:</u> This follows from Theorem 1.1.2 and (1.3.9).

<u>Theorem 1.3.7:</u> Let h be a kernel of degree m satisfying (1.3.8) and $\zeta_m < \infty$. Then

$$n^{-1/2}(V_n(h) - \vartheta) \longrightarrow N(0,m^2\zeta_1)$$

weakly as $n \longrightarrow \infty$.

<u>Proof:</u> Theorem 1.3.1 and (1.3.9).

Theorem 1.3.8: Let h be a kernel of degree m satisfying $\zeta_m < \infty$, (1.3. 8) and $\zeta_1 > 0$. Define

$$Y_n(t) = \begin{cases} (nm^2\zeta_1)^{-1/2}\ k(V_k(h) - \vartheta) & \text{if } t = k/n \\ \text{by linear interpolation elsewhere} \end{cases}$$

and $Z_n(t) = Y_n([nt]/n)$. Then Y_n and Z_n $(n \geq 1)$ converge weakly to the standard Wiener process.

Proof: This follows immediately from Theorem 1.3.2 and (1.3.10) and since

$$\max_{0 \leq k \leq m} (nm^2\zeta_1)^{-1/2}\ k(V_k(h) - \vartheta) \longrightarrow 0 \qquad \text{a.s.}$$

Theorem 1.3.9: Let h be a kernel as in the previous theorem. Then there exists (w.l.o.g., cf. Theorem 1.3.4) a standard Brownian motion $\{B(t) : t \geq 0\}$ such that with probability one

$$\lim_{t\to\infty} (t \log\log t)^{-1/2}\ |\ \zeta^{-1/2} m^{-1}\ t(V_{[t]}(h) - \vartheta) - B(t)| \rightarrow 0.$$

Theorem 1.3.10: Let h be as before. Then

$$\overline{\lim_{n\to\infty}}\ (n \log\log n)^{-1/2}\ n\ (\dot{V}_n(h) - \vartheta) = m\,\zeta_1^{1/2} \qquad \text{a.s.}$$

4. Generalized U-statistics

Three years after Hoeffding published the first paper on U-statistics, Lehmann generalized the concept to a multi-sample version. Since some statistics - which are important for practical purposes - fall into this class, we shall discuss the additional aspects appearing in this connection. Many of the things previously said carry over. Therefore the presentation will be kept rather short and - for simplicity in notation - only the two-sample case will be considered.

Definition 1.4.1: Let $(E,\mathcal{B})$ be a measurable space and let $h : E^{m_1+m_2} \longrightarrow \mathbb{R}$ $(m_1, m_2 \in \mathbb{N},\ m = m_1+m_2)$ be a measurable map satisfying the following symmetry property: $h(x,y) = h(\tau x, \sigma y)$ for all $x \in E^{m_1}$, $y \in E^{m_2}$, $\tau \in \gamma_{m_1}$ and $\sigma \in \gamma_{m_2}$. Such an h will be called a *generalized kernel of degree* (m_1,m_2).

A generalized kernel h of degree (m_1,m_2) estimates the parameter

$$\vartheta = \vartheta(F,G) = \iint h(x,y)\, dF^{m_1}(x)\, dG^{m_2}(y)$$

for $F,G \in M(\mathcal{B})$, provided the integral exists. The symmetry condition in 1.4.1 is, as in the 1-sample case, not essential, since for an arbitrary measurable map $h : E^m \to \mathbb{R}$ the function

$$h(x,y) = m_1!^{-1} m_2!^{-1} \sum_{\tau \in \gamma_{m_1}} \sum_{\sigma \in \gamma_{m_2}} h(\tau x, \sigma y)$$

is symmetric in the above sense and estimates the same parameter.

Definition 1.4.2:

Let $h : E^m \longrightarrow \mathbb{R}$ be a kernel of degree (m_1, m_2). A map

$$(1.4.1) \qquad U_{n_1,n_2} = U_{n_1,n_2}(h) = \binom{n_1}{m_1}^{-1} \binom{n_2}{m_2}^{-1} \sum_{\pi_1 \in \Pi_{n_1}} \sum_{\pi_2 \in \Pi_{n_2}} h(\pi_1 x, \pi_2 y),$$

defined for $x \in E^{n_1}$ and $y \in E^{n_2}$ where $n_1 \geq m_1$ and $n_2 \geq m_2$, is called a *generalized U-statistic*, more precisely a *two-sample U-statistic*. Here, Π_{n_i} denotes the set of all projections from E^{n_i} onto m_i coordinates $1 \leq j_1 < j_2 < \ldots j_{m_i} \leq n_i$ $(i=1,2)$.

Similar to this definition we obtain a *generalized V-statistic* by setting

$$V_{n_1,n_2} = V_{n_1,n_2}(h) = n_1^{-m_1} n_2^{-m_2}$$

$$(1.4.2) \qquad \sum_{1 \leq i_1,\ldots,i_{m_1} \leq n_1} \sum_{1 \leq j_1,\ldots,j_{m_2} \leq n_2} h(x_{i_1},\ldots,x_{i_{m_1}}, y_{j_1},\ldots,y_{j_{m_2}})$$

where $x = (x_1,\ldots,x_{n_1}) \in E^{n_1}$ and $y = (y_1,\ldots,y_{n_2}) \in E^{n_2}$. Again, we shall make no distinction between the consideration of a generalized U-statistic or a V-statistic as a map defined on $E^{n_1+n_2}$ or regarding them to be random variables in case the probabilities F and G are fixed.

Example 1.4.1:

The Wilcoxon two-sample test for difference in location under symmetry assumption is probably the best-known two-sample U-statistic. Let $X_1,\dots,X_{n_1}$, $Y_1,\dots,Y_{n_2}$ be independent random variables. The X_i are supposed to have a continuous d.f. F and the Y_j a continuous d.f. G. Now consider the combined sample $X_{in} = X_i$ $(i \leq n_1)$ and $X_{in} = Y_{i-n_1}$ $(n_1 < i \leq n_1+n_2=n)$. Denote by R_i $(1 \leq i \leq n)$ the rank of X_{in} among $X_{1n},\dots,X_{nn}$ and define the *Wilcoxon two-sample statistic* by

$$(1.4.3) \qquad W = \sum_{i=1}^{n_1} R_i .$$

This is in fact a U-statistic since

$$W = \sum_{i=1}^{n_1} \sum_{j=1}^{n_2} 1_{\{X_i - Y_j \geq 0\}} + \frac{n_1(n_1+1)}{2} .$$

Here the kernel is given by $h : \mathbb{R}^2 \to \mathbb{R}$, $h(x,y) = 1_{\{y \leq x\}}$ and it estimates $\vartheta = \vartheta(F,G) = \int G(x)dF(x)$. Under H_o, i.e. $F = G$, $\vartheta = 1/2$ and hence

$$EW = 1/2(n_1 n_2 + n_1(n_1+1)) = 1/2\, n_1(n+1)$$

and

$$\operatorname{Var} W = \sum_{i,j=1}^{n_1} \sum_{k,l=1}^{n_2} P(Y_k \leq X_i, Y_l \leq X_j) - 1/4\, n_1^2 n_2^2$$

$$= 1/12\, n_1 n_2 (n+1).$$

For practical purposes this test should be applied only if F can be assumed to be symmetric.

Example 1.4.2:

Another test for location alternatives was introduced by Lehmann. Let $X_1,\dots,X_{n_1}$, $Y_1,\dots,Y_{n_2}$ be independent random variables as in the previous example. The kernel is given by

$$h(x_1,x_2,y_1,y_2) = \begin{cases} 1 & \text{if } \begin{array}{l}\min(x_1,x_2)>\max(y_1,y_2) \text{ or} \\ \min(y_1,y_2)>\max(x_1,x_2).\end{array} \\ 0 & \text{otherwise}. \end{cases}$$

If F and G are absolutely continuous, the corresponding U-statistics provide good tests. The estimated parameter can be derived as follows: Setting $A = \{\min(x_1,x_2) > \max(y_1,y_2)\}$, we have

$$F^2\times G^2(A) = \int\int G^2(x_1 \wedge x_2)dF(x_1)dF(x_2) =$$

$$= 2\int G^2(x) \ (1-F(x))dF(x)$$

and similarly

$$F^2\times G^2(A) = \int\int (1-F(y_1 \vee y_2))^2 \ dG(y_1)dG(y_2) =$$

$$= 2\int (1-F(y))^2 \ G(y) \ dG(y).$$

Therefore

$$\vartheta = \vartheta(F,G) = \int G^2(x)(1-F(x))dF(x) +$$

$$+ \int (1-G(x))^2 \ F(x) \ dF(x)$$

$$+ \int F^2(y)(1-G(y))dG(y) +$$

$$+ \int (1-F(y))^2 G(y)dG(y) =$$

$$= 1/3 + \int (F(x) - G(x))^2 \ d(F(x) + G(x)),$$

and if $F = G$, $\vartheta = 1/3$.

Example 1.4.3:

Let $X_1,\ldots,X_{n_1}$, $Y_1,\ldots,Y_{n_2}$ be as before, but F and G need not be continuous. Assume that F and G have finite expectations. In order to estimate a difference in the expectations one can use the two-sample U-statistic with kernel $h(x,y) = x-y$ of degree (1,1). Then

$$U_{n_1,n_2}(h) = n_1^{-1} n_2^{-2} \sum_{i,j} X_i - Y_j = \overline{X} - \overline{Y}.$$

More generally, if the one-sample U-statistic $U_{n_1}(h)$ estimates $\vartheta(F)$, then $U_{n_1}(h) - U_{n_2}(h)$ estimates $\vartheta(F) - \vartheta(G)$ and is a 2-sample U-statistic.

Example 1.4.4:

Let $X_1,\ldots,X_{n_1}$, $Y_1,\ldots,Y_{n_2}$ be as in Example 1.4.1. Assume that

$F(x) = H(\frac{x-\vartheta_1}{\eta_1})$ and $G(x) = H(\frac{x-\vartheta_2}{\eta_2})$, where ϑ_1 and ϑ_2 are unknown and where the test problem is specified by $H_o = \{\eta_1=\eta_2\}$ against $H_1=\{\eta_1<\eta_2\}$. (The Y_j's are more spread-out than the X_i's!). Lehmann suggested to use the kernel

$$h(x_1,x_2,y_1,y_2) = \begin{cases} 1 & \text{if } |y_1-y_2| - |x_1-x_2| > 0 \\ 0 & \text{otherwise}. \end{cases}$$

The corresponding 2-sample U-statistic can be viewed as the scale analogon to the Wilcoxon 2-sample statistic in Example 1.4.1! Since under H_o Y_1-Y_2 has the same distribution as X_1-X_2, we obtain $\vartheta = 1/2$.

Two-sample U-statistics have the same optimality properties as discussed in section 1 in the one-sample case (cf. Theorem 1.1.1).We shall not discuss this property further, and turn to a.s. convergence properties and weak convergence directly. We first note the following, already used in the previous examples:

Lemma 1.4.1:

A two-sample U-statistic with kernel h of degree (m_1,m_2) is an unbiased estimator of $\vartheta(F,G)$.

Definition 1.4.3:

A kernel h of degree (m_1,m_2) is called *degenerate* w.r. to F and G in $M(\mathcal{B})$, if for all $x_1,\ldots,x_{m_1},y_1,\ldots$ $\ldots,y_{m_2} \in E$

$$(1.4.4)\quad \left[\begin{array}{l} \int h(u,x_2,\ldots,x_{m_1},y_1,\ldots,y_{m_2})\,dF(u) = 0 \\ \text{and} \\ \int h(x_1,\ldots,x_{m_1},u,y_2,\ldots,y_{m_2})\,dG(u) = 0. \end{array}\right.$$

Definition 1.4.4:

A field $\{Z_{u,v} : u,v \geq 1\}$ of real-valued random variables $Z_{u,v}$ $(u,v\geq 1)$ is called a * *martingale* with respect to the filtration $\{\mathcal{B}_{r,s} : r,s\geq 1\}$ of σ-algebras $\mathcal{B}_{r,s}$ if

$$E(Z_{u,v}\,|\,\mathcal{B}_{r,s}) = Z_{k,l}$$

where $k = \min(u,r)$ and $l=\min(v,s)$, for all $u,v,r,s \geq 1$.

For such a *-martingale, $\xi_u := \sup_{n_2\leq v\leq N_2} |Z_{u,v}|$

$(n_2,N_2$ fixed, $u\geq 1)$ is a submartingale. Indeed,

$$E(\xi_u\,|\,\mathcal{B}_{u-1,N_2}) \geq \sup_{n_2\leq v\leq N_2} \left|E(Z_{u,v}\,|\,\mathcal{B}_{u-1,N_2})\right|$$

$$= \sup_{n_2\leq v\leq N_2} |Z_{u-1,v}| = \xi_{u-1}.$$

The martingale properties of a two-sample U-statistic are summarized in

Lemma 1.4.2:

Let h be a generalized kernel of degree (m_1,m_2) and let $\{X_k : k\geq 1\}$ and $\{Y_l : l\geq 1\}$ be two independent sequences of i.i.d. random variables with d.f. F and G respectively. Assume that $\int\int|h(x,y)|\,dF^{m_1}(x)dG^{m_2}(y) < \infty$, and define the σ-fields $\mathcal{A}_{r,s}$ and $\mathcal{F}_{r,s}$ $(r,s\leq 1)$ by

$$\mathcal{A}_{r,s} = \sigma(X_1,\ldots,X_r,Y_1,\ldots,Y_s)$$

and

$$\mathcal{F}_{r,s} = \sigma(X^{(r)}, X_{r+1}, \ldots; Y^{(s)}, Y_{s+1}, \ldots) \quad \text{(cf. sect. 1).}$$

Then $\{U_{n_1,n_2}(h) : n_1 \geq m_1, n_2 \geq m_2\}$ is a backward *-martingale w.r. to $\{\mathcal{F}_{r,s} : r \geq m_1, s \geq m_2\}$, and, moreover, if h is degenerate, then

$\{\binom{n_1}{m_1}\binom{n_2}{m_2} U_{n_1,n_2}(h) : n_1 \geq m_1,\ n_2 \geq m_2\}$ is a *-martingale w.r. to

$\{A_{r,s} : r \geq m_1, s \geq m_2\}$.

Proof:

Let us first note that

$$U_{n_1,n_2}(h) = E(h(X_1, \ldots, X_{m_1}, Y_1, \ldots, Y_{m_2}) \,|\, (X^{(n_1)}, Y^{(n_2)})).$$

Indeed, $U_{n_1,n_2}(h)$ is $\sigma(X^{(n_1)}, Y^{(n_2)})$ - measurable and if A belongs to this σ-algebra, then

$$\int_A h(X_1, \ldots, X_{m_1}, Y_1, \ldots, Y_{m_2})\,dP =$$

$$= \binom{n_1}{m_1}^{-1} \binom{n_2}{m_2}^{-1} \int_A \sum_{1 \leq i_1 < \ldots < i_{m_1} \leq n_1} \; \sum_{1 \leq j_1 < \ldots < j_{m_2} \leq n_2} h(X_{i_1}, \ldots, X_{i_{m_1}}, Y_{j_1}, \ldots, Y_{j_{m_2}})\,dP$$

$$= \int_A U_{n_1,n_2}(h)\,dP.$$

Using this fact, we can write

$$E(U_{n_1,n_2}(h) \,|\, \mathcal{F}_{r,s}) =$$

$$= E(E(h(X_1, \ldots, X_{m_1}, Y_1, \ldots, Y_{m_2}) \,|\, X^{(n_1)}, Y^{(n_2)}) \,|\, \mathcal{F}_{r,s}) =$$

$$= E(E(h(X_1, \ldots, X_{m_1}, Y_1, \ldots, Y_{m_2}) \,|\, \mathcal{F}_{n_1,n_2}) \,|\, \mathcal{F}_{r,s}) =$$

$$= E(h(X_1, \ldots, X_{m_1}, Y_1, \ldots, Y_{m_2}) \,|\, \mathcal{F}_{n_1 \vee r, n_2 \vee s})$$

$$= U_{r \vee n_1, s \vee n_2}(h),$$

since $\mathcal{F}_{r,s} \cap \mathcal{F}_{n_1,n_2} = \mathcal{F}_{r\vee n_1, s\vee n_2}$.

Now let h be degenerate. Then

$$E(\binom{n_1}{m_1}\binom{n_2}{m_2}U_{n_1,n_2}(h) \mid \mathcal{A}_{r,s}) =$$

$$= \sum_{1\leq i_1<..<i_{m_1}\leq n_1} \sum_{1\leq j_1<..<j_{m_2}\leq n_2} E(h(X_{i_1},\ldots,X_{i_{m_1}}, Y_{j_1},\ldots,Y_{j_{m_2}}) \mid \mathcal{A}_{r,s})$$

$$= \binom{n_1\wedge r}{m_1}\binom{n_2\wedge s}{m_2} U_{n_1\wedge r, n_2\wedge s}(h),$$

since, e.g., if $i_{m_1} > r$, then $E(h(\ldots X_{i_{m_1}},\ldots) \mid \mathcal{A}_{r,s}) = \int h(\ldots,u,\ldots)dF(u) = 0$ by degeneracy and symmetry.

<u>Proposition 1.4.1:</u>
Let h be a degenerate generalized kernel of degree (m_1,m_2). If

(1.4.5) $\quad \|h\|^2 := E\, h^2(X_1,\ldots,X_{m_1},Y_1,\ldots,Y_{m_2}) < \infty,$

then

$$\operatorname{Var} U_{n_1,n_2}(h) = \binom{n_1}{m_1}^{-1}\binom{n_2}{m_2}^{-1} \|h\|^2 .$$

<u>Proof:</u>
Since $Eh(X_1,\ldots,X_{m_1}, Y_1,\ldots,Y_{m_2}) = 0$,

$$\operatorname{Var} U_{n_1,n_2}(h) =$$

$$= \binom{n_1}{m_1}^{-2}\binom{n_2}{m_2}^{-2} \sum Eh(X_{i_1},\ldots,X_{i_{m_1}}, Y_{j_1},\ldots,Y_{j_{m_2}}) \cdot h(X_{k_1},\ldots,X_{k_{m_1}},Y_{l_1},\ldots,Y_{l_{m_2}}).$$

Each summand equals zero unless $j_u = l_u$ and $i_v = k_v$ for all $1\leq u\leq m_2$ and $1\leq v\leq m_1$, because h is degenerate and the random variables are independent.

Theorem 1.4.1:
Let h be a generalized kernel of degree (m_1,m_2), and let $\{X_k : k \geq 1\}$ and $\{Y_l : l \geq 1\}$ be two independent sequences of i.i.d. random elements with distributions F and G respectively. Then

$$\lim_{\substack{n_1\to\infty \\ n_2\to\infty}} U_{n_1,n_2}(h) = \vartheta(F,G) \quad \text{a.s. and in } L_1,$$

provided $h \in L \log_1 L$.
This means that

$$C(h) := \int\int |h(x,y)| \, (\log^+ |h(x,y)|) \, dF^{m_1}(x) dG^{m_2}(y) < \infty.$$

Proof:
We first prove the a.s. convergence if $h \in L_2$, i.e., if (1.4.5) holds.

Since for $n_i < N_i$ $(i=1,2)$ $\{ \sup_{n_2 \leq j \leq N_2} |U_{k,j}(h)-\vartheta| : n_1 \leq k \leq N_1\}$ is a nonnegative reversed submartingale by Lemma 1.4.2 and the remark after Definition 1.4.4., it follows from Doob's inequality (Doob, 1953, Theorem 3.2, p.314) that

$$P\Big(\sup_{\substack{n_1 \leq k \leq N_1 \\ n_2 \leq j \leq N_2}} |U_{k,j}(h)-\vartheta| \geq \varepsilon \Big) \leq \varepsilon^{-2} E \sup_{n_2 \leq j \leq N_2} |U_{n_1,j}(h)-\vartheta|^2 .$$

Also note that by Lemma 1.1.1 $\{U_{n_1,j}(h)-\vartheta : n_2 \leq j \leq N_2\}$ is a reversed martingale (n_1 is fixed!) and so by Doob's inequality (Doob, 1953, Theorem 3.4, p.317) we have

$$P(\sup |U_{k,j}(h)-\vartheta| \geq \varepsilon) \leq 4\varepsilon^{-2} E(U_{n_1,n_2}(h)-\vartheta)^2.$$

In order to prove the a.s. convergence if $h \in L_2$ it now suffices to show that $U_{n_1,n_2}(h) - \vartheta \to 0$ in L_2. But this can be shown similarly to the argument in the previous proof. Write

$$E(U_{n_1,n_2}(h) - \vartheta)^2 =$$

$$= \binom{n_1}{m_1}^{-2} \binom{n_2}{m_2}^{-2} \sum E(h(X_{i_1},\ldots,X_{i_{m_1}},Y_{j_1},\ldots,Y_{j_{m_2}})-\vartheta) \cdot$$

$$\cdot (h(X_{k_1},\ldots,X_{k_{m_1}},Y_{l_1},\ldots,Y_{l_{m_2}}) - \vartheta)$$

where the summation extends over all $1 \leq i_1 < \ldots < i_{m_1} \leq n_1$, $1 \leq j_1 < \ldots < j_{m_2} \leq n_2$, $1 \leq k_1 < \ldots < k_{m_1} \leq n_1$ and $1 \leq l_1 < \ldots < l_{m_2} \leq n_2$. If all i_u are different from the $k_{u'}$'s and if all j_v are different from all $l_{v'}$'s, then the expectation under the sum is zero. In all other cases this expectation can be bounded by $2\|h\|^2$, using Hölder's inequality. The number of these cases is exactly

$$\binom{n_1}{m_1}^2 \binom{n_2}{m_2}^2 - \frac{n_1! \qquad n_2!}{(m_1!)^2 (m_2!)^2 (n_1-2m_1)!(n_2-2m_2)!} =$$

$$= \binom{n_1}{m_1}^2 \binom{n_2}{m_2}^2 \left(1 - \frac{(n_1-m_1)\ldots(n_1-2m_1+1)}{n_1\ldots(n_1-m_1+1)} \frac{(n_2-m_2)\ldots(n_2-2m_2+1)}{n_2\ldots(n_2-m_2+1)}\right)$$

$$= O\left(\binom{n_1}{m_1}^2 \binom{n_2}{m_2}^2 (n_1 \wedge n_2)^{-1}\right),$$

consequently

$$E(U_{n_1,n_2}(h) - \vartheta)^2 = O((n_1 \wedge n_2)^{-1} \|h\|^2),$$

where O does not depend on h, F and G. This completes the proof of the a.s. convergence in case $h \in L_2$.

Now let h be an arbitrary kernel in $L \log_1 L$.

Given $\eta > 0$, we can find a kernel $\tilde{h} \in L_2$ of degree (m_1,m_2) satisfying

$$\iint |h(x,y) - \tilde{h}(x,y)| \log^+ |h(x,y) - \tilde{h}(x,y)| \, dF^{m_1}(x) dG^{m_2}(y) < \eta.$$

Denoting the estimable parameter associated with $\tilde{h}$ by $\tilde{\vartheta}$, it follows from Doob's inequalities again for any constant K that

$$P\Big(\sup_{\substack{n_1 \leq k \leq N_1 \\ n_2 \leq j \leq N_2}} |U_{k,j}(h-\tilde{h}) - \vartheta + \tilde{\vartheta}| \geq \varepsilon \Big) \leq$$

$$\leq \varepsilon^{-1} E \sup_{n_2 \leq j \leq N_2} |U_{n_1,j}(h-\tilde{h}) - \vartheta + \tilde{\vartheta}| \leq$$

$$\leq \varepsilon^{-1} \Big[\frac{e}{K(e-1)} + \frac{e}{e-1} E \big|U_{n_1,n_2}(h-\tilde{h}) - \vartheta + \tilde{\vartheta}\big| \log^+ K \big|U_{n_1,n_2}(h-\tilde{h}) - \vartheta + \tilde{\vartheta}\big| \Big].$$

Choosing $\log K = \eta^{-1/2}$, $n_1 = m_1$ and $n_2 = m_2$ we conclude that

$$P(\sup_{k,j} |U_{k,j}(h-\tilde{h}) - \vartheta + \tilde{\vartheta}| \geq \varepsilon) = O(\varepsilon^{-1}\eta^{1/2}),$$

and hence

$$\limsup_{\substack{n_1 \to \infty \\ n_2 \to \infty}} |U_{n_1,n_2}(h) - \vartheta - U_{n_1,n_2}(\tilde{h}) + \tilde{\vartheta}|$$

$$= \limsup_{\substack{n_1 \to \infty \\ n_2 \to \infty}} |U_{n_1,n_2}(h) - \vartheta| \leq \varepsilon$$

with probability $> 1 - O(\varepsilon^{-1}\eta^{1/2})$. If $\varepsilon \to 0$ and if $\eta = \eta(\varepsilon) = \varepsilon^3$, the a.s. convergence follows.

In order to prove the L_1-convergence, note that $U_{k,j}(h)$ is a backward *-martingale w.r.t. $\mathcal{F}_{r,s}$ by Lemma 1.4.2. Therefore, if $k \geq m_1$ and $j \geq m_2$,

$$E(|U_{k,j}(h)| \log^+ |U_{k,j}(h)|) = E(|E(h|\mathcal{F}_{k,j})| \log^+ |E(h|\mathcal{F}_{k,j})|) \leq$$

$$\leq E(E(|h| \log^+ |h| \,|\, \mathcal{F}_{k,j})) = E(|h| \log^+ |h|)$$

where h stands for $U_{m_1,m_2}(h)$, and for $K > 0$

$$E(1_{\{|U_{k,j}(h)| \geq K\}} |U_{k,j}(h)|) \leq (\log K)^{-1} E(|U_{k,j}(h)| \log^+ |U_{k,j}|$$

$$\leq (\log K)^{-1} E(|U_{k,j}(h)| \log^+ |U_{k,j}(h)|) \leq \frac{C(h)}{\log K}.$$

Hence $\{U_{n_1,n_2}(h)-\vartheta : n_1 \geq m_1, n_2 \geq m_2\}$ is uniformly integrable and we obtain the L_1-convergence from the a.s. convergence.

The Decomposition Theorem 1.2.1 certainly has its analogue in the present case. Let h be a generalized kernel of degree (m_1,m_2). For $0 \leq c \leq m_1$ and $0 \leq d \leq m_2$ define

$$(1.4.6) \quad \tilde{h}_{cd}(x_1,\ldots,x_c, y_1,\ldots,y_d) =$$

$$= Eh(x_1,\ldots,x_c,X_{c+1},\ldots,X_{m_1},y_1,\ldots,y_d,Y_{d+1},\ldots,Y_{m_2})$$

and

$$(1.4.7) \quad h_{cd}(x,y) = \sum_{k=o}^{c} \sum_{l=o}^{d} (-1)^{c+d-k-l} \sum_{\substack{K\subset\{1,..,c\}\\ |K|=k}} \sum_{\substack{L\subset\{1,..,d\}\\ |L|=l}} \tilde{h}_{kl}((x)_K,(y)_L)$$

where $x = (x_1,\ldots,x_c)$, $y = (y_1,\ldots,y_d)$, where $(x)_K$ denotes the k-tupel consisting of those coordinates of x belonging to K and where $(y)_L$ denotes the l-tupel of those coordinates of y belonging to L. Note that $\tilde{h}_{oo} = h_{oo} = \vartheta(F,G)$, $\tilde{h}_{m_1m_2} = h$, $\tilde{h}_{cd}$ and h_{cd} are symmetric and (exercise) h_{cd} is a degenerate generalized kernel of degree (c,d) if $c,d > 0$. If $c = o$ and $d > 1$ (or $c > 1$ and $d = 0$) h_{cd} is a degenerate kernel of degree d (resp. c).

<u>Theorem 1.4.2:</u> (Decomposition Theorem for two-sample U-statistics)

Let h be a generalized kernel of degree (m_1,m_2). Then

$$(1.4.8) \quad U_{n_1,n_2}(h) = \sum_{c=o}^{m_1} \sum_{d=o}^{m_2} \binom{m_1}{c}\binom{m_2}{d} U_{n_1,n_2}(h_{cd}).$$

<u>Proof:</u>

There is not much difference to the proof of Theorem 1.2.1. First observe that

$$h(x,y) = \sum_{c=o}^{m_1} \sum_{d=o}^{m_2} \sum_{\substack{K\subset\{1,..,m_1\}\\ |K|=c}} \sum_{\substack{L\subset\{1,..,m_2\}\\ |L|=d}} h_{cd}((x)_K,(y)_L)$$

since

$$\sum_{k=o}^{m_1-i}\sum_{l=o}^{m_2-j}(-1)^{k+l}\binom{m_1-i}{k}\binom{m_2-j}{l}=\begin{cases}1 & \text{if } i=m_1 \text{ and } j=m_2\\ 0 & \text{otherwise.}\end{cases}$$

Thus we can write

$$U_{n_1,n_2}(h)=\binom{n_1}{m_1}^{-1}\binom{n_2}{m_2}^{-1}\sum_{1\leq i_1<..<i_{m_1}\leq n_1}\ \sum_{1\leq j_1<..<j_{m_2}\leq n_2}\ \sum_{c=o}^{m_1}\sum_{d=o}^{m_2}$$

$$\sum_{1\leq k_1<..<k_c\leq m_1}\ \sum_{1\leq l_1<..<l_d\leq m_2} h_{cd}(X_{i_{k_1}},\ldots,X_{i_{k_c}},Y_{j_{l_1}},\ldots,Y_{j_{l_d}}).$$

If $i_{k_1},\ldots,i_{k_c}$, $j_{l_1},\ldots,j_{l_d}$ are fixed then there exist exactly $\binom{n_1-c}{m_1-c}\binom{n_2-d}{m_2-d}$ choices of different indices $i_{k_{c+1}},\ldots,i_{k_{m_1}}$, $j_{l_{d+1}},\ldots,j_{l_{m_2}}$. Hence it follows that

$$U_{n_1,n_2}(h)=\sum_{c=o}^{m_1}\sum_{d=o}^{m_2}\binom{n_1-c}{m_1-c}\binom{n_2-d}{m_2-d}\binom{n_1}{m_1}^{-1}\binom{n_2}{m_2}^{-1}$$

$$\sum_{1\leq i_1<..<i_c\leq n_1}\ \sum_{1\leq j_1<..<j_d\leq n_2} h_{cd}(X_{i_1},\ldots,X_{i_c},Y_{j_1},\ldots,Y_{j_d}).$$

The theorem follows from the formula $\binom{n-c}{m-c}\binom{n}{m}^{-1}\binom{n}{c}=\binom{m}{c}$.

The main result to be proved in this section is an invariance principle in $D([0,1]^2)$. This space is a direct generalization of $D([0,1])$ and contains all functions f defined on $[0,1]^2$ with values in $\mathbb{R}$ satisfying the following conditions:

(i) For all $0 \leq s,t \leq 1$ the limits of the following sequences exist if $s_n \to s$ and $t_n \to t$:

$\{f(s_n,t_n) : s_n < s,\ t_n < t\}$, $\{f(s_n,t_n) : s_n \geq s,\ t_n < t\}$

$\{f(s_n,t_n) : s_n < s,\ t_n \geq t\}$, $\{f(s_n,t_n) : s_n \geq s,\ t_n \geq t\}$.

(ii) For the last sequence in (i) the limit equals $f(s,t)$.

For $f,g \in D([0,1]^2)$ we can obtain a metric d_2 by

defining $d_2(f,g)$ to be the infimum of those numbers ε for which there exist $\lambda,\mu : [0,1] \to [0,1]$ surjective, strictly increasing and continuous such that $\sup_t |\lambda(t)-t| < \varepsilon$, $\sup_s |\mu(s)-s| < \varepsilon$ and $\sup_{s,t} |f(s,t)-g(\mu(s),\lambda(t))| < \varepsilon$. It is easily seen that this defines a metric and clearly $(D([0,1]^2), d_2)$ is then a separable metric space containing $C([0,1]^2) = \{f : [0,1]^2 \to \mathbb{R} \text{ continuous}\}$ as a closed subspace. Note that $d_2(f,g) \leq \|f-g\|_\infty$ and that both metrics induce the same topology on $C([0,1]^2)$. On $D([0,1]^2)$ we shall consider the Borel-σ-field generated by d_2, so that its restriction to $C([0,1]^2)$ coincides with the Borel-σ-field generated by the sup-metric. We shall not need criteria for weak convergence in $D([0,1]^2)$ but the following

Lemma 1.4.3:

The map $\psi : D([0,1])^2 \to D([0,1]^2)$ defined by $\psi(f,g)(s,t) = tf(s) + sg(t)$ is continuous.

Proof:

Let $\lambda,\mu : [0,1] \to [0,1]$ be strictly increasing, surjective and continuous maps. Then

$$|\psi(f,g)(s,t) - \psi(f',g')(\lambda(s),\mu(t))| \leq$$

$$\leq |f(s) - f'(\lambda(s))| + |g(t) - g'(\mu(t))|$$

shows that $d_2(\psi(f,g), \psi(f',g')) \leq 2\max\{d(f,f'), d(g,g')\}$ where d denotes the Skorohod metric on $D([0,1])$. This proves the continuity of ψ, because the right-hand side is twice the metric on $D([0,1])^2$.

Theorem 1.4.3:

Let h be a generalized kernel of degree (m_1,m_2). Define the $D([0,1]^2)$-valued random elements Z_n by

$$(1.4.9) \qquad Z_n(s,t) = \begin{cases} [sn][tn]n^{-3/2}(U_{[sn],[tn]}(h)-\vartheta) & \text{if } m_2 \leq tn \leq n, m_1 \leq sn \leq n \\ 0 & \text{otherwise.} \end{cases}$$

Then $\{Z_n : n \geq 1\}$ converges weakly in $D([0,1]^2)$ to the distribution of the random function $(s,t) \to t\zeta_{10}^{1/2} m_1 W_1(s) + s\zeta_{01}^{1/2} m_2 W_2(t)$, where W_1 and W_2 are two independent standard Wiener processes and where $\zeta_{10} = \operatorname{Var} h_{10}$ and $\zeta_{01} = \operatorname{Var} h_{01}$.

Proof:

In the decomposition formula (1.4.8) the summands for $c=0$, $d=1$ and $c=1$, $d=0$ are sums of i.i.d. random variables:

$$m_1 n_1^{-1} \sum_{i=1}^{n_1} h_{10}(X_i) \quad \text{and} \quad m_2 n_2^{-1} \sum_{j=1}^{n_2} h_{01}(Y_j).$$

By Donsker's theorem (Billingsley, 1968, p.137)

$$Z_{1n}(s) = n^{-1/2} \zeta_{10}^{-1/2} \sum_{i=1}^{[ns]} h_{10}(X_i) \to W_1(s)$$

and

$$Z_{2n}(t) = n^{-1/2} \zeta_{01}^{-1/2} \sum_{j=1}^{[nt]} h_{01}(Y_j) \to W_1(t)$$

weakly in $D([0,1])$. Since Z_{1n} and Z_{2n} are independent, it follows from Theorem 3.2 in Billingsley (1968, p.21), that $(Z_{1n}, Z_{2n}) \to (W_1, W_2)$ in $D([0,1])^2$, where here W_1 and W_2 are independent standard Wiener processes, naturally embedded into $D([0,1])^2$, equivalently $(m_1\zeta_{10}^{1/2} Z_{1n}, m_2\zeta_{01}^{1/2} Z_{2n}) \to (m_1\zeta_{10}^{1/2} W_1, m_2\zeta_{01}^{1/2} W_2)$. Since the map ψ from Lemma 1.4.3 is continuous, it follows from Theorem 5.1 in Billingsley (1968, p.30) that $\psi(m_1\zeta_{10}^{1/2} Z_{1n}, m_2\zeta_{01}^{1/2} Z_{2n}) \to \psi(m_1\zeta_{10}^{1/2} W_1, m_2\zeta_{01}^{1/2} W_2)$ weakly in $D([0,1]^2)$.

It is left to show that (using the Decomposition Theorem 1.4.2)

$$Z_n(s,t) - \psi(m_1\zeta_{10}^{1/2} Z_{1n}, m_2\zeta_{01}^{1/2} Z_{2n})(s,t) =$$

$$(1.4.10) \quad = (u(s,t)-t) n^{-\frac{1}{2}} m_1 \sum_{i=1}^{[ns]} h_{10}(X_i) + (v(s,t)-s) n^{-\frac{1}{2}} m_2 \sum_{j=1}^{[nt]} h_{01}(Y_j)$$

$$+ [tn][sn] n^{-3/2} \sum_{c=1}^{m_1} \sum_{d=1}^{m_2} \binom{m_1}{c} \binom{m_2}{d} U_{[ns][nt]}(h_{cd})$$

converges to zero in probability w.r. to the sup-metric, where

$u(s,t) = n^{-1}[nt]$, $v(s,t) = n^{-1}[ns]$ if $m_1 \leqq ns$ and $m_2 \leqq nt$ and where $u=v=0$ otherwise.

Fix $c,d \geqq 1$. Using Proposition 1.4.1., Chow's inequality (Lemma 2.3.1) and Doob's inequality (Doob, 1953, Theorem 3.4, p.317) we obtain

$$P(\sup_{m_1 \leqq k \leqq n} k\, n^{-3/2} \sup_{m_2 \leqq j \leqq n} j\, |U_{k,j}(h_{cd})| \geqq \varepsilon) \leqq$$

$$\leqq P(\sup_{m_1 \leqq k \leqq n} k\, n^{-1} \sup_{m_2 \leqq j \leqq n^{1/2}} |U_{k,j}(h_{cd})| \geqq \varepsilon) +$$

$$+ P(\sup_{m_1 \leqq k \leqq n} k\, n^{-1/2} \sup_{n^{1/2} \leqq j \leqq n} |U_{k,j}(h_{cd})| \geqq \varepsilon) \leqq$$

$$\leqq \varepsilon^{-2} n^{-2} \sum_{k=m_1+1}^{n} (2k-1)\, E \sup_{m_2 \leqq j \leqq n^{1/2}} U_{k,j}(h_{cd})^2 +$$

$$+ \varepsilon^{-2} n^{-2} m_1^2\, E \sup_{m_2 \leqq j \leqq n^{1/2}} U_{m_1,j}(h_{cd})^2 +$$

$$+ \varepsilon^{-2} n^{-1} \sum_{k=m_1+1}^{n} (2k-1)\, E \sup_{n^{1/2} \leqq j \leqq n} U_{k,j}(h_{cd})^2 +$$

$$+ \varepsilon^{-2} n^{-1} m_1^2\, E \sup_{n^{1/2} \leqq j \leqq n} U_{m_1,j}(h_{cd})^2 \leqq$$

$$\leqq \varepsilon^{-2} n^{-2} \sum_{k=m_1+1}^{n} (2k-1) \binom{k}{c}^{-1} \binom{m_2}{d}^{-1} + \varepsilon^{-2} n^{-2} m_1^2 \binom{m_1}{c}^{-1} \binom{m_2}{d}^{-1} +$$

$$+ \varepsilon^{-2} n^{-1} \sum_{k=m_1+1}^{n} (2k-1) \binom{k}{c}^{-1} \binom{[\sqrt{n}]+1}{d}^{-1} + \varepsilon^{-2} n^{-1} m_1^2 \binom{m_1}{c}^{-1} \binom{[\sqrt{n}]+1}{d}$$

$$= O(n^{-1/2}).$$

Obviously, $\sup\{n^{-1/2}\, t \sum_{i=1}^{[ns]} h_{10}(X_i) + n^{-1/2} s \sum_{j=1}^{[nt]} h_{01}(Y_j) :$

$s \leqq \frac{m_1}{n}$ or $t \leqq \frac{m_2}{n}\} \to 0$ in probability and since

$\left|\frac{[nr]}{n} - r\right| \leqq n^{-1}$ for all $0 \leqq r \leqq 1$, one easily shows that the first two summands in (1.4.10) also converge to zero in probability w.r. to the sup-norm (Exercise). This completes the proof of the theorem.

Corollary:

Let h be a generalized kernel of degree (m_1,m_2). Then $\sqrt{n_1+n_2}\,(U_{n_1,n_2}(h)-\vartheta)$ converges weakly to $N(0,\sigma^2)$ as $n_1,n_2 \to \infty$ in such a way that $n_1(n_1+n_2)^{-1} \to \lambda \in (0,1)$, where $\sigma^2 = \lambda^{-1}\zeta_{10}\, m_1^2 + (1-\lambda)^{-1}\zeta_{01}\, m_2^2$.

Proof:

For n_1,n_2 define $g_{n_1,n_2} : D([0,1]^2) \to \mathbb{R}$ by

$g_{n_1,n_2}(f) = f(\frac{n_1}{n_1+n_2}, \frac{n_2}{n_1+n_2})$. Then $g_{n_1,n_2}(f_n) \longrightarrow g(f)$ (as $n_1,n_2 \longrightarrow \infty$ in the above sense and $n \longrightarrow \infty$) for every sequence $f_n \in D([0,1]^2)$ with $f_n \longrightarrow f \in C([0,1]^2)$, where $g(f) = f(\lambda,1-\lambda)$. Hence

$$g_{n_1,n_2}(Z_{n_1+n_2}) \to g(\psi(m_1\, \zeta_{10}^{1/2} W_1, m_2\, \zeta_{01}^{1/2} W_2))$$

weakly by Theorem 5.5 in Billingsley (1968, p.34) and by the continuity of the limit process in Theorem 1.4.3. The corollary follows observing that

$$g_{n_1,n_2}(Z_{n_1+n_2}) = \frac{n_1}{n_1+n_2}\,\frac{n_2}{n_1+n_2}\,\sqrt{n_1+n_2}\,(U_{n_1,n_2}(h)-\vartheta),$$

$$g(\psi(m_1\, \zeta_{10}^{1/2} W_1, m_2\, \zeta_{01}^{1/2} W_2)) =$$

$$= (1-\lambda)\zeta_{10}^{1/2} m_1 W_1(\lambda) + \lambda\zeta_{01}^{1/2} m_2 W_2(1-\lambda)$$

and that the distribution of this last variable is $N(0,\tau^2)$ with $\tau^2 = (1-\lambda)^2\lambda m_1^2\, \zeta_{10} + \lambda^2(1-\lambda) m_2^2\, \zeta_{01}$.

Example 1.4.5:

For the Wilcoxon two-sample statistic (Example 1.4.1) we have $\tilde{h}_{10}(x) = G(x)$ and $\tilde{h}_{01}(y) = 1-F(y)$. Under H_o (i.e. $F=G$) it follows $\zeta_{10} = \zeta_{01} = 1/12$ so that the asymptotic variance becomes $\sigma^2 = (12\lambda(1-\lambda))^{-1}$.

For Lehmann's statistic in Example 1.4.2 we have

$$\tilde{h}_{10}(x) = \iiint 1_{\{x \wedge x_2 > y_1 \vee y_2\}} \, dG(y_1)dG(y_2)dF(x_2) +$$

$$+\iiint 1_{\{y_1 \wedge y_2 > x \vee x_2\}} \, dG(y_1)dG(y_2)dF(x_2)$$

$$= \int G(x_2)^2 dF(x_2) - 2 \int_x^\infty G(x_2)dF(x_2) +$$

$$+ 1 + G(x)^2 - 2F(x)G(x).$$

Under H_o (i.e. F=G) it follows that

$$\tilde{h}_{10}(x) = 4/3 - 2 \int_x^\infty F(y)dF(y) - F(x)^2 = 1/3.$$

Similarly, $\tilde{h}_{01}(x) = \tilde{h}_{10}(x)$. It follows that in this case the asymptotic variance is $\sigma^2 = 0$, and hence $\sqrt{n_1+n_2}\,(U_{n_1,n_2}(h) - 1/3) \to 0$ in probability. In other words, the kernel is degenerate. In the next chapter we shall discuss the method how to derive limit laws for properly normalized U-statistics when the kernel is degenerate.

We add one more result concluding this section. The proof of it is very similar to the proof of Proposition 1.3.2 and therefore left as an exercise. It determines the asymptotic distribution of generalized V-statistics.

Proposition 1.4.2:

Let h be a generalized kernel of degree (m_1,m_2) satisfying

$$\max\{E(h(X_{i_1},\ldots,X_{i_{m_1}},Y_{j_1},\ldots,Y_{j_{m_2}}))^2 : 1 \leq j_k \leq m_2, 1 \leq i_l \leq m_1\} < \infty.$$

Then

$$\lim_{n\to\infty} \|[sn][tn]n^{-3/2}(U_{[sn],[tn]}(h) - V_{[sn],[tn]}(h))\|_\infty = 0$$

in probability, where V denotes the generalized V-statistic from (1.4.2). Moreover,

$$\lim_{n\to\infty} \sup_{k,j} kj\, n^{-3/2} |\vartheta - E\, V_{k,j}(h)| = 0.$$

Notes on chapter 1:

The decomposition method discussed in section 2 is due to Hoeffding (1948) as well as the asymptotic normality in section 3. In the generalized case this was done by Lehmann (1951). The backward martingale property goes back to Berk (1966) and the invariance principles have been stated by Miller and Sen (1972, 1974). The law of the iterated logarithm is due to Serfling. The book of Randles and Wolfe (1979) contains an extensive discussion of examples and applications and Serfling's book (1980) contains further results on U- and V-statistics in the non-degenerate case.

Chapter 2: Differentiable statistical functionals

The asymptotic theory of nonlinear statistical functionals can be developed by some differential calculus, in the sense of Fréchet, Hadamard (compact) or 'Gateaux' derivatives. This idea was von Mises' approach in 1947. The first non-vanishing term in its expansion determines the type of the asymptotic distribution in general; if it is the first order term the limit will be normal, in all other cases different types of distributions arise which can be described as multiple stochastic integrals with respect to the Brownian bridge (Filippova 1962). In particular, since the V-statistics of chapter 1 fall into this category, the results of the preceding chapter will be extended to U- and V-statistics in the degenerate case.

Von Mises' work in 1947 did not have much influence on the development in statistics until the late sixties when this technique became important in robust statistics.

1. Definition of differentiable statistical functionals

For two probability measures (or distribution functions) $F^{(i)}$ $(i = 0,1)$ on $\mathbb{R}$ define $F^{(t)} = (1-t)\, F^{(0)} + t\, F^{(1)}$ $(0 \leq t \leq 1)$. A set A of probability measures is called $F^{(0)}$*-star shaped* if $F^{(t)} \in A$ whenever $F^{(1)} \in A$ and $0 \leq t \leq 1$.

Definition 2.1.1: Let $T : U \longrightarrow \mathbb{R}$ be a functional defined on a subset $U \subset M(\mathcal{B}_1)$ of probability measures on $\mathbb{R}$. T is called *m-times differentiable* in $F^{(0)} \in U$ with respect to the $F^{(0)}$-star shaped set $A \subset U$ if the following two conditions hold:

(2.1.1) For each $F^{(1)} \in A$ the map $t \longrightarrow T(F^{(t)})$ $(0 \leq t \leq 1)$ is m-times differentiable in $t = 0$. (Differentiability at the endpoints of the interval [0,1] is understood in an obvious manner.)

(2.1.2) For each $p = 1,\ldots,m$ there exists a symmetric, measurable map $\varphi^{(p)} : \mathbb{R}^p \longrightarrow \mathbb{R}$ satisfying

$$\frac{d^p}{dt^p} T(F^{(t)}) \Big|_{t=0} = \int \ldots \int \varphi^{(p)}(y_1,\ldots,y_p) \prod_{i=1}^{p} d(F^{(1)}-F^{(0)})(y_i).$$

If a functional T(F) is m+1 - times differentiable in the above sense and if the empirical d.f. $F^{(1)} = F_n$ based on n i.i.d. observations with distribution $F = F^{(0)}$ belongs to A a.s., then by Taylor's theorem

$$(2.1.3) \qquad T(F_n) - T(F) = \sum_{k=1}^{m} \frac{1}{k!} \frac{d^k}{dt^k} T(F^{(t)}) \Big|_{t=0} + \\ + \frac{1}{(m+1)!} \frac{d^{m+1}}{dt^{m+1}} T(F^{(t)}) \Big|_{t=t_0}$$

where $0 \leq t_0 \leq 1$ and hence the asymptotic behaviour is determined by V-statistics provided the remainder term tends to zero in probability. This motivates the following two definitions.

Definition 2.1.2: A functional $T : U \longrightarrow \mathbb{R}$ is called a *differentiable statistical functional of order m* in $F = F^{(0)} \in U$ if there exists an F-star shaped $A \subset U$ such that the following conditions are satisfied:

(2.1.4) For the empirical d.f. F_n of n independent, F-distributed random variables one has

$$\lim_{n \to \infty} P(F_n \in A) = 1.$$

(2.1.5) T is m-times differentiable in F with respect to A.

Definition 2.1.3: A functional $T : U \longrightarrow \mathbb{R}$ is called a *von Mises' functional of order m* in $F = F^{(0)} \in U$ if it is a differentiable statistical functional of order m+1 in F and if:

(2.1.6) For all $\delta, \varepsilon > 0$ and all $p = 1,\ldots,m+1$

$$\lim_{n \to \infty} P\left(n^{p/2 - \delta} \sup_{0 \leq s \leq 1} \left| \frac{d^p}{dt^p} T(F_n^{(t)}) \Big|_{t=s} \right| > \varepsilon \right) = 0$$

where $F_n^{(t)} = tF_n + (1-t)F$.

Suppose that $1 \leq k \leq m$ is the smallest integer such that

$\frac{d^k}{dt^k} T(F_n^{(t)}) \Big|_{t=0}$ does not vanish. Then (2.1.6) implies that the remaining terms in the Taylor expansion (2.1.3) tend to zero in probability, even after being blown up by $n^{k/2}$. However, the properties of a von Mises' functional are too strong to be satisfied for many statistics. Let k be as above and define

$$\text{(2.1.7)} \qquad \text{Rem}_k(T,F) = T(F_n) - T(F) - \frac{1}{k!}\frac{d^k}{dt^k}T(F_n^{(t)})\Big|_{t=0} .$$

In many examples we shall show that $n^{k/2}\ \text{Rem}_k(T,F) \longrightarrow 0$ (as $n \to \infty$) in probability.

Stronger notions of differentiability than condition (2.1.1) imply that $n^{1/2}\ \text{Rem}_1(T,F) \longrightarrow 0$ in probability. Two results are given here.

Proposition 2.1.1: Let $T : U \longrightarrow \mathbb{R}$ be a functional defined on an open subset $U \subset C_b(\mathbb{R})$, the space of bounded real valued functions on $\mathbb{R}$ equipped with the supremum-norm $\| \ \|_\infty$. If T is Fréchet differentiable at the continuous d.f. $F \in U$, then

$$n^{1/2}\ \text{Rem}_1(T,F) \longrightarrow 0 \qquad (\text{as } n \to \infty)$$

in probability.

Proof: Recall that T is *Fréchet differentiable* at F if there exists a linear functional T' (on the Banach space $C_b(\mathbb{R})$) such that

$$\text{(2.1.8)} \qquad \lim_{\|G-F\|_\infty \to 0} \frac{|T(G) - T(F) - T'(G-F)|}{\|G-F\|_\infty} = 0.$$

Since the Kolmogorov-Smirnov statistic $\|F_n - F\|_\infty$ satisfies $n^{1/2}\ \|F_n - F\|_\infty = O(1)$ in probability, the proposition follows immediately from (2.1.8) setting $G = F_n$.

Proposition 2.1.2: Let $T : U \longrightarrow \mathbb{R}$ be a functional defined on an open subset $U \subset D([0,1])$ where $D([0,1])$ is considered as the complete, non-separable Banach space with the supremum-norm. If T is Hadamard differentiable at the uniform distribution $F \in U$ then

$$\lim_{n \to \infty} n^{1/2}\ \text{Rem}_1(T,F) = 0$$

in probability.

Proof: Recall that T is called *Hadamard differentiable* at F if there exists a linear functional T' such that for any compact subset $K \subset D([0,1])$

$$\text{(2.1.9)} \qquad \lim_{t \to 0} \sup_{G \in K} \frac{T(F+tG) - T(F) - T'(tG)}{t} = 0.$$

Let $Y_1, Y_2, \dots$ be independent, uniformly distributed random variables. Denote by F_n the empirical d.f. of $Y_1,\dots,Y_n$ and by F_n^* the continuous d.f. having uniform distribution on each of the intervals $[Y_{(i-1)}, Y_{(i)}]$, $[0, Y_{(1)}]$ and $[Y_{(n)}, 1]$ $(1 < i \leq n)$. Since by Donsker's theorem $n^{1/2}(F_n^* - F)$ converges weakly in $C([0,1])$ to the Brownian bridge $\{B(t) : 0 \leq t \leq 1\}$, the sequence

$$P_n = \mathcal{L}(n^{1/2}(F_n^* - F)) \qquad (n \geq 1)$$

is relatively compact and hence by Prohorov's theorem it is also tight.

Let $\varepsilon > 0$ be given. Then there exists a compact set $C \subset C([0,1])$ with $P_n(C) \geq 1 - \varepsilon$ $(n \geq 1)$. Denote by

$$d(f,A) = \inf \{ \|f-g\|_\infty : g \in A\}$$

the distance of $f \in D([0,1])$ and $A \subset D([0,1])$ and consider now C to be a compact set in $(D([0,1]), \|\ \|_\infty)$. Since $\|F_n^* - F_n\|_\infty \leq n^{-1}$ it follows that

$$\{d(n^{1/2}(F_n - F), C) \leq n^{-1/2}\} \supset \{n^{1/2}(F_n^* - F) \in C\} = E_n .$$

Thus $P(E_n) \geq 1 - \varepsilon$.

The proposition will follow if we can show that for large n

$$|n^{1/2} \operatorname{Rem}_1(T,F)| < \varepsilon \quad \text{on } E_n.$$

In order to prove this assume that the statement is false. Then there exists a sequence $n_k \longrightarrow \infty$ with

$$|n_k^{1/2} \operatorname{Rem}_1(T,F)| \geq \varepsilon \qquad (k \geq 1)$$

and

$$d(n_k^{1/2}(F_{n_k} - F), C) \leq n_k^{-1/2} \qquad (k \geq 1).$$

Choose $H_k \in C$ with $\|n_k^{1/2}(F_{n_k} - F) - H_k\|_\infty \leq n_k^{-1/2}$. Since C is compact we may assume that $\|H_k - H\|_\infty \longrightarrow 0$ for some $H \in C$ and therefore we have $\|n_k^{1/2}(F_{n_k} - F) - H\|_\infty \longrightarrow 0$. Thus $\{ n_k^{1/2}(F_{n_k} - F), H : k \geq 1 \}$ is compact and (2.1.9) implies that there exists a k_o such that

$$n_k^{1/2} |T(F_{n_k}) - T(F) - T'(F_{n_k} - F)| = |n_k^{1/2} \operatorname{Rem}_1(T,F)| < \varepsilon.$$

for all $k \geq k_o$. This contradicts the assumption.

We shall now discuss a few examples in the remaining part of this section. Some basic facts from calculus are not explicitly stated but are obvious in each case. For example, conditions for interchanging differentiation and integration, the fact that $\int f'(x)\,dx = 0$ for a differentiable density f etc. We shall not give all computations in each example.

Example 2.1.1: Let $T(F) = \int x\,dF(x)$ be defined for all F where the integral exists. T is called the *empirical expectation*. This set clearly is star shaped and

$$T(F^{(t)}) = \int x\,dF(x) + t\int x\,d(F^{(1)}-F)(x).$$

Therefore

$$T(F_n) - T(F) = \int x\,d(F_n-F)(x).$$

More generally, let us consider the *sample central k-th moment*

$$T(F) = \int_{-\infty}^{\infty}\Big(x - \int_{-\infty}^{\infty} y\,dF(y)\Big)^k dF(x)\ .$$

Here, the map to be considered is given by

$$T(F^{(t)}) = \int_{-\infty}^{\infty}\Big(x - \mu_F - t(\mu_G-\mu_F)\Big)^k dF^{(t)}(x)$$

where $F = F^{(0)}$, $G = F^{(1)}$, $\mu_F = \int x\,dF(x)$ and $\mu_G = \int x\,dG(x)$. This map is C^∞, the first derivative is

$$\frac{d}{dt}T(F^{(t)})\Big|_{t=s} = -\int_{-\infty}^{\infty} k\Big(x - \mu_F - s(\mu_G-\mu_F)\Big)^{k-1}(\mu_G-\mu_F)\,dF^{(s)}(x) + \int_{-\infty}^{\infty}\Big(x - \mu_F - s(\mu_G-\mu_F)\Big)^k d(G-F)(x)$$

and the second derivative is

$$\frac{d^2}{dt^2}T(F^{(t)})\Big|_{t=s} = k(k-1)(\mu_G-\mu_F)^2\int_{-\infty}^{\infty}\Big(x-\mu_F-s(\mu_G-\mu_F)\Big)^{k-2} dF^{(s)}(x) - 2k(\mu_G-\mu_F)\int_{-\infty}^{\infty}\Big(x-\mu_F-s(\mu_G-\mu_F)\Big)^{k-1} d(G-F)(x).$$

Setting $s = 0$ it follows that (2.1.2) is satisfied with the (symmetric) maps

$$x \longrightarrow (x - \mu_F)^k - k\Big(\int_{-\infty}^{\infty}(y-\mu_F)^{k-1}\,dF(y)\Big)\,x$$

and

$$(x,y) \longrightarrow k(k-1)\left(\int_{-\infty}^{\infty} (z-\mu_F)^{k-2} \, dF(z) \right) xy - 2kx(y-\mu_F)^{k-1} .$$

We have shown that T is a differentiable statistical functional. In order that T is a von Mises' functional of order 1 we have to check (2.1.6). We leave this as an exercise: Both derivatives can be written as a sum of V-statistics with degenerate kernels. Then use the fact that for such a statistic with degenerate kernel h of degree p (cf. Proposition 2.2.2 below)

$$E\left(n^{p/2-\delta} \, V_n(h) \right)^2 \longrightarrow 0.$$

Example 2.1.2: Many differentiable statistical functionals are defined implicitly, for example by some minimizing condition. One of them is the *maximum likelihood estimator (ML-estimator)*. Let $\Theta \subset \mathbb{R}$ be open and let $\{ F(\cdot,\vartheta) : \vartheta \in \Theta \}$ be a family of d.f. $F(\cdot,\vartheta)$ with densities $f(\cdot,\vartheta)$. We shall make the following assumptions:

(2.1.10a) For all $x \in \mathbb{R}$ the map $\vartheta \longrightarrow f(x,\vartheta)$ is three times differentiable.

(2.1.10b) For each $\vartheta_o \in \Theta$, the absolute values of the three derivatives

$$\frac{\partial f(x,\vartheta)}{\partial \vartheta} , \quad \frac{\partial^2 f(x,\vartheta)}{\partial \vartheta^2} \quad \text{and} \quad \frac{\partial^3 \log f(x,\vartheta)}{\partial \vartheta^3}$$

are uniformly bounded in some neighbourhood of ϑ_o by functions $u(x)$, $v(x)$ and $w(x)$ respectively satisfying

$$\int u(x) \, dx < \infty , \quad \int v(x) \, dx < \infty \quad \text{and} \quad \int w(x) \, dF(x) < \infty .$$

(2.1.10c) $$0 < \int \left(\left. \frac{\partial \log f(x,\vartheta)}{\partial \vartheta} \right|_{\vartheta=\vartheta_o} \right)^2 dF(x,\vartheta_o) < \infty$$

for every $\vartheta_o \in \Theta$.

(2.1.10d) $$0 < \int \left(\left. \frac{\partial f(x,\vartheta)}{\partial \vartheta} \right|_{\vartheta=\vartheta_o} \right)^2 (f(x,\vartheta_o))^{-1} \, dx < \infty$$

for every $\vartheta_o \in \Theta$.

The maximum likelihood estimator of ϑ is defined by the solution of

$$\sum_{i=1}^{n} \frac{\partial \log f(X_i,\vartheta)}{\partial\vartheta} = 0 \quad \text{for which} \quad \sum_{i=1}^{n} \frac{\partial^2 \log f(X_i,\vartheta)}{\partial\vartheta^2} < 0.$$

(Of course, this solution should be uniquely determined or the ML-estimator is defined to be just one solution of the equation.)

We see that the ML-estimator is given by a functional $T(F)$ satisfying the equation $\int g(x,T(F))\, dF(x) = 0$ where

$$g(x,\vartheta') = \left.\frac{\partial \log f(x,\vartheta)}{\partial\vartheta}\right|_{\vartheta=\vartheta'} .$$

Define $g'(x,\vartheta') = \left.\frac{\partial g(x,\vartheta)}{\partial\vartheta}\right|_{\vartheta=\vartheta'}$ and

$$g''(x,\vartheta') = \left.\frac{\partial^2 g(x,\vartheta)}{\partial\vartheta^2}\right|_{\vartheta=\vartheta'} .$$

We show first that a solution exists for $F = F(\cdot,\vartheta_o)$ and $tF_n + (1-t)F$, uniformly in $t \in [0,1]$, if n is sufficiently large.

In fact by (2.1.10b) $\int g(x,\vartheta_o)\, dF(x,\vartheta_o) = 0$. Therefore $T(F(\cdot,\vartheta_o)) = \vartheta_o$ is a solution. Let $F^{(1)} = F_n$ denote the empirical d.f. of a sample of size n drawn from the distribution $F^{(0)} = F$.

Let $0 < t \leq 1$ be fixed. By (2.1.10b,d)

$$\frac{\partial}{\partial\vartheta}\left[\int g(x,\vartheta)\, dF(x)\right]\Big|_{\vartheta=\vartheta_o} = -\int \left(\left.\frac{\partial f(x,\vartheta)}{\partial\vartheta}\right|_{\vartheta=\vartheta_o}\right)^2 \frac{1}{f(x,\vartheta_o)}dx$$

$$< 0$$

and hence $\int g(x,\vartheta)\, dF(x) < 0$ if $\vartheta > \vartheta_o$ and > 0 if $\vartheta < \vartheta_o$. Since $F_n \longrightarrow F$ a.s. a similar relation holds for F_n instead of F, if n is sufficiently large. Consequently they also must hold for any convex combination $F^{(t)}$ where $0 \leq t \leq 1$. Condition (2.1.4) is easily derived from this.

Next we show (2.1.5) for the map $t \longrightarrow T(F^{(t)})$, if there exists a solution for F_n. We keep the same notation as before. Since $\int g(x,T(F^{(t)}))\, dF^{(t)}(x) \equiv 0$ $(0 \leq t \leq 1)$ we have

$$0 = \frac{d}{dt}T(F^{(t)})\Big|_{t=s} \int g'(x,T(F^{(s)}))\, dF^{(s)}(x) +$$

$$+ \int g(x,T(F^{(s)}))\, d(F_n-F)(x) .$$

Therefore (2.1.1) and (2.1.2) are satisfied, since
$\int g'(x,T(F))\,dF(x) \neq 0$.
Finally we show (2.1.6), i.e. that T is a von Mises' functional of order 0. Setting $b(t) = \int g'(x,T(F^{(t)}))\,dF^{(t)}(x)$ we observe that

$$|b(t)| \geq |\int g'(x,\vartheta_o)\,dF(x)| - \int |g'(x,T(F^{(t)}))-g'(x,\vartheta_o)|\,dF^{(t)}(x) - |\int g'(x,\vartheta_o)\,d(F_n-F)(x)| .$$

Since

$$\lim_{n\to\infty} \sup_{0\leq t\leq 1} T(F^{(t)}) = T(F) = \vartheta_o \text{ in probability,}$$

$$|g'(x,T(F^{(t)}))-g'(x,T(F))| \leq w(x)\,|T(F^{(t)})-\vartheta_o|$$

and since $g'(\ ,\vartheta_o) \in L_1(F)$, for $\eta > 0$ there exists an $n_o \in \mathbb{N}$ satisfying

$$P(\inf_{0\leq t\leq 1} |b(t)| \geq 2^{-1}|\int g'(x,\vartheta_o)\,dF(x)| > 0) \geq 1 - \eta$$

for all $n \geq n_o$. Similarly, with probability $\geq 1-\eta$ we have for $n \geq n_o$ (say) that

$$\sup_{0\leq t\leq 1} |\int_{-\infty}^{\infty} g'(x,T(F^{(t)}))\,d(F_n-F)(x)| < \varepsilon$$

where $\varepsilon > 0$. It follows that for $n \geq n_o$ with probability $\geq 1 - 2\eta$

$$\left|\sup_{0\leq t\leq 1} \frac{d}{ds}T(F^{(s)})\Big|_{s=t}\right| \leq 2\left|\int g'(x,\vartheta_o)\,dF(x)\right|^{-1} \left(|\int_{-\infty}^{\infty} g(x,\vartheta_o)\,d(F_n-F)(x)| + \varepsilon|\sup_{0\leq t\leq 1} \frac{d}{ds}T(F^{(s)})\Big|_{s=t}|\right)$$

equivalently [with $\varepsilon' \longrightarrow 0$ as $\varepsilon \longrightarrow 0$]

$$\left|\sup_{0\leq t\leq 1} \frac{d}{ds}T(F^{(t)})\Big|_{s=t}\right| \leq 2(1-\varepsilon')^{-1}|\int_{-\infty}^{\infty} g'(x,\vartheta_o)\,dF(x)|^{-1} |\int_{-\infty}^{\infty} g(x,\vartheta_o)\,d(F_n-F)(x)| .$$

Since $g(\cdot,\vartheta_o) \in L_2(F)$,

$$n^{1/2} \int g(x,\vartheta_o)\,d(F_n-F)(x)$$

is stochastically bounded. (2.1.6) follows. The details are left as an exercise.

In the discrete case the ML-estimator is defined by the solution of the equation $H(t,\alpha) = 0$ where

$$H(t,\alpha) = \sum_{j=1}^{r} \frac{p_j'(\alpha)}{p_j(\alpha)} \int 1_{(a_j,a_{j+1}]}(x)\, dF_n^{(t)}(x).$$

Here $F_n^{(t)} = F + t(F_n - F)$, $F = F(\cdot,\alpha)$, $p_j(\alpha) = F(a_{j+1}) - F(a_j)$, $p_j' = \frac{\partial p_j}{\partial \alpha}$ and $-\infty = a_1 < a_2 < \ldots < a_{r+1} = \infty$. We leave it as an exercise to show that $T(F)$ is again a von Mises' functional of order 0 in $F(\cdot,\alpha)$ for each $\alpha \in \mathbb{R}$, provided a condition similar to (2.1.10) holds. Also the reader may check that in both cases $T(F)$ is a von Mises' functional of order 1 under suitable assumptions.

Example 2.1.3: We shall use the same notation as in Example 2.1.2 for the discrete case. Define $n_j = n(F_n(a_{j+1}) - F_n(a_j))$ where F_n denotes the empirical d.f. given by n i.i.d. observations. The χ^2-*minimum estimator* minimizes

$$\sum_{j=1}^{r} \frac{(n_j - np_j(\alpha))^2}{np_j(\alpha)} = \left(n^{-1} \sum_{j=1}^{r} \frac{n_j^2}{p_j(\alpha)} \right) - n.$$

Since p_j is differentiable by assumption, this value of α can be obtained from the requirement that the derivative vanishes,

$$\sum_{j=1}^{r} \frac{n_j^2\, p_j'(\alpha)}{n\, p_j^2(\alpha)} = 0,$$

and that the second derivative (if it exists) is negative. Setting

$$h(\alpha,x,y) = \begin{cases} \dfrac{p_j'(\alpha)}{p_j^2(\alpha)} & \text{if } a_j < x,y \leq a_{j+1} \\ 0 & \text{otherwise} \end{cases}$$

it follows immediately that

$$\int\int h(\alpha,x,y)\, dF_n(x)\, dF_n(y) = n^{-2} \sum_{j=1}^{r} \frac{p_j'(\alpha)}{p_j^2(\alpha)}\, n_j^2 ,$$

and therefore we may define the χ^2-minimum estimator $T(F)$ by

$$\int\int h(T(F),x,y)\, dF(x)\, dF(y) = 0.$$

In this example we shall make the same assumptions as in the previous one. (Note that in (2.1.10) we have to replace f by

$p_j(\alpha) \quad (1 \leq j \leq r)$.

Since $\int\int h(\alpha,x,y)\ dF(x,\alpha)\ dF(y,\alpha) = \sum_{1\leq j\leq r} p_j'(\alpha) = 0$

and

$$\frac{\partial}{\partial\vartheta}\left[\int\int h(\vartheta,x,y)\ dF(x,\alpha)\ dF(y,\alpha)\right]\Bigg|_{\vartheta=\alpha} =$$

$$= -2 \sum_{j=1}^{r} (p_j(\alpha))^{-1} (p_j'(\alpha))^2 < 0,$$

(2.1.4) is verified as in Example 2.1.2.

The differentiability of $t \longrightarrow T(F^{(t)})$ also follows as before. Implicit differentiation yields (using the symmetry of h)

$$0 = \frac{d}{dt} \int\int h(T(F^{(t)}),x,y)\ dF^{(t)}(x)\ dF^{(t)}(y) =$$

$$= \frac{d}{dt}\Big\{ \int\int h(T(F^{(t)}),x,y)\ dF(x)\ dF(y) +$$

$$+ t \int\int h(T(F^{(t)}),x,y)\ [dF(x)d(F_n-F)(y)+dF(y)d(F_n-F)(x)]$$

$$+ t^2 \int\int h(T(F^{(t)}),x,y)\ d(F_n-F)(x)\ d(F_n-F)(y)\Big\} =$$

$$= \frac{d}{dt}T(F^{(t)})\Big\{ \int\int h'(T(F^{(t)}),x,y)\ dF(x)\ dF(y) +$$

$$+ 2t \int\int h'(T(F^{(t)}),x,y)\ dF(x)\ d(F_n-F)(y) +$$

$$+ t^2 \int\int h'(T(F^{(t)}),x,y)\ d(F_n-F)(x)\ d(F_n-F)(y)\Big\} +$$

$$+ 2 \int\int h(T(F^{(t)}),x,y)\ dF(x)\ d(F_n-F)(y) +$$

$$+ 2t \int\int h(T(F^{(t)}),x,y)\ d(F_n-F)(x)\ d(F_n-F)(y)\ .$$

Therefore

$$\frac{d}{dt}T(F^{(t)})\Big|_{t=0} = -2\left(\int\int h'(T(F),x,y)\ dF(x)\ dF(y)\right)^{-1}$$

$$\int\int h(T(F),x,y)\ dF(x)\ d(F_n-F)(y)$$

where $\int\int h'(T(F),x,y)\, dF(x)\, dF(y) < 0$ as shown before and where h' denotes the partial derivative with respect to the first coordinate. This shows that T is a differentiable statistical functional of order 1.
Condition (2.1.6) can be shown similarly to Example 2.1.2 (Exercise).

Example 2.1.4: (Example 2.1.3 continued)
The χ^2-*statistic* is defined by

$$\chi^2 = n^{-1} \left(\sum_{j=1}^{r} \frac{n_j^2}{p_j(\alpha)} \right) - n$$

where α is fixed. Recall that

$$n_j^2 = n^2 \int\int 1_{(a_j,a_{j+1}]}(x)\, 1_{(a_j,a_{j+1}]}(y)\, dF_n(x)\, dF_n(y) \quad .$$

Thus χ^2 can be written as

$$\chi^2 = n \left\{ \int\int h(x,y)\, dF_n(x)\, dF_n(y) - 1 \right\}$$

where $h(x,y) = p_j(\alpha)^{-1}$ if $a_j < x,y \leqq a_{j+1}$ and $h(x,y) = 0$ otherwise. The corresponding statistical functional T is defined by $T(G) = n(\int\int h(x,y)\, dG(x)\, dG(y) - 1)$ with derivatives

$$\frac{d}{dt}T(F^{(t)})\bigg|_{t=s} = 2sn \int\int h(x,y)\, d(F_n-F)(x) d(F_n-F)(y)$$

and

$$\frac{d^2}{dt^2}T(F^{(t)})\bigg|_{t=s} = 2n \int\int h(x,y)\, d(F_n-F)(x)\, d(F_n-F)(y),$$

since for $a_j < y \leqq a_{j+1}$

$$\int h(x,y)\, dF(x) = p_j(\alpha)^{-1} \int 1_{(a_j,a_{j+1}]}(x)\, dF(x) = 1$$

and therefore $\int\int h(x,y)\, dF(x)\, d(F_n-F)(y) = 0$.
We have shown that T is a differentiable statistical functional of order 2 at $F = F(\cdot,\alpha)$.
Note that α was known in the preceding arguments. If α is unknown one can replace α by some estimator S. Let us assume that S itself is a differentiable statistical functional of order $m \geqq 2$ at F satisfying $S(F) = \alpha$. Then we have to consider

$$T(G) = n \sum_{j=1}^{r} \iint p_j(S(G))^{-1} 1_{(a_j,a_{j+1}]}(x)\, 1_{(a_j,a_{j+1}]}(y)\, dG(x)\, dG(y) - n.$$

Let $g_j(x,y) = 1$ if $a_j < x,y \leq a_{j+1}$ and $g_j(x,y) = 0$ otherwise $(1 \leq j \leq r)$. With this notation

$$n^{-1} \frac{d}{dt}T(F^{(t)})\Big|_{t=s} = \sum_{j=1}^{r} - \frac{p_j'(S(F^{(s)}))}{p_j^2(S(F^{(s)}))} \frac{d}{dt}S(F^{(t)})\Big|_{t=s} \iint g_j(x,y)\, dF^{(s)}(x)\, dF^{(s)}(y) +$$

$$+ \sum_{j=1}^{r} \frac{2}{p_j(S(F^{(s)}))} \iint g_j(x,y)\, dF(x)\, d(F_n-F)(y) +$$

$$+ \sum_{j=1}^{r} \frac{2s}{p_j(S(F^{(s)}))} \iint g_j(x,y)\, d(F_n-F)(x)\, d(F_n-F)(y)$$

and

$$n^{-1} \frac{d^2}{dt^2}T(F^{(t)})\Big|_{t=s} = \sum_{j=1}^{r} -p_j^{-3}(S)\Big[p_j(S)\left\{p_j''(S)S'^2+p_j'(S)S''\right\}$$

$$- 2\, S'^2\, (p_j'(S))^2\Big] \iint g_j(x,y)\, dF^{(s)}(x)\, dF^{(s)}(y) -$$

$$- 4 \sum_{j=1}^{r} \frac{p_j'(S)\, S'}{p_j^2(S)} \iint g_j(x,y)\, dF^{(s)}(x)\, d(F_n-F)(y) +$$

$$+ 2 \sum_{j=1}^{r} p_j^{-1}(S) \iint g_j(x,y)\, d(F_n-F)(x)\, d(F_n-F)(y),$$

where we used

$$S = S(F^{(s)}), \quad S' = \frac{d}{dt}S(F^{(t)})\Big|_{t=s} \text{ and } S'' = \frac{d^2}{dt^2}S(F^{(t)})\Big|_{s}.$$

Setting s=0 we see that

$$\frac{d}{dt}T(F^{(t)})\Big|_{t=0} = 0$$

and

$$n^{-1} \frac{d^2}{dt^2}T(F^{(t)})\Big|_{t=0} = \sum_{j=1}^{r} 2\, p_j^{-1}(S(F))\, p_j'(S(F))^2\left(\frac{d}{dt}S(F^{(t)})\Big|_0\right)^2$$

$$- 4 \sum_{j=1}^{r} p_j^{-1}(S(F))\, p_j'(S(F))\, \frac{d}{dt}S(F^{(t)})\Big|_0 \int 1_{(a_j,a_{j+1}]}(y)\, d(F_n-F)(y) +$$

$$+ 2 \sum_{j=1}^{r} p_j^{-2}(S(F)) \iint g_j(x,y)\, d(F_n-F)(x)\, d(F_n-F)(y) .$$

Thus T(F) is a differentiable statistical functional.
Finally we remark that both functionals considered in this example are von Mises' functionals, in the latter case we have to assume in addition that S is a von Mises' functional of order $m \geq 1$. This can be shown similarly to the previous examples using estimates of the second moments of double stochastic integrals. (Exercise)

Example 2.1.5: The ω^2-*minimum estimator* T(F) is derived from the Cramér - Smirnov - von Mises test statistic in a similar way as the χ^2-minimum estimator was derived from the χ^2-test statistic. It minimizes

$$\omega(\alpha) = \int (F(x,\alpha) - F_n(x))^2 \, dF(x,\alpha),$$

and it is implicitly defined by the equations

$$\frac{\partial}{\partial\alpha}\left[\int (F(x,\alpha) - G(x))^2 \, dF(x,\alpha)\right]\Big|_{\alpha=T(G)} = 0$$

and

$$\frac{\partial^2}{\partial\alpha^2}\left[\int (F(x,\alpha) - G(x))^2 \, dF(x,\alpha)\right]\Big|_{\alpha=T(G)} > 0,$$

if suitable differentiability conditions are made.
Assume that $\alpha \longrightarrow F(x,\alpha)$ is C^3 $(x \in \mathbb{R})$ in a neighbourhood of ϑ_o and that the derivatives are uniformly bounded by $L_2(dx)$ functions. Moreover, suppose that $F(\cdot,\alpha)$ has a twice differentiable density (in some neighbourhood of ϑ_o), also bounded by some square integrable function . Finally we assume that

$$0 < \int \left(\frac{\partial F(x,\alpha)}{\partial\alpha}\Big|_{\alpha=\vartheta_o} \right)^2 dF(x,\vartheta_o) < \infty .$$

Certainly, setting $F = F(\cdot,\vartheta_o)$, then $T(F) = \vartheta_o$ and

$$\frac{\partial^2}{\partial\alpha^2}\left[\int (F(x,\alpha)-F(x))^2 \, dF(x,\alpha)\right]\Big|_{\alpha=\vartheta_o} =$$

$$= 2\int \left(\frac{\partial F(x,\alpha)}{\partial\alpha}\Big|_{\alpha=\vartheta_o} \right)^2 dF(x,\vartheta_o) > 0.$$

As before it can be shown that $T(F_.^{(t)})$ is well defined with large probability. It is also left as an exercise to show that T is a differentiable statistical functional and even a von Mises' functional at F.

The generalized Cramér-von Mises statistic is defined by

$$\int w(F)\ (F_n - F)^2\ dF$$

where F is some known d.f. and where w is a known function. The functional is defined by

$$T(G) = \int w(F)\ (G - F)^2\ dF$$

and has derivatives

$$\frac{d}{dt}T(F^{(t)})\Big|_{t=s} = 2 \int w(F)\ (F^{(s)}-F)\ (F_n-F)\ dF$$

and

$$\frac{d^2}{dt^2}T(F^{(t)})\Big|_{t=s} = 2 \int w(F)\ (F_n-F)^2\ dF =$$

$$= 2 \int\int\int 1_{\{x \geq \max(u,v)\}} w(F(x))dF(x)\ d(F_n-F)(u)d(F_n-F)(v)\ .$$

Therefore T is a von Mises' functional of order 2 at F.

Example 2.1.6: Let $\{F(\cdot,\vartheta)\ :\ \vartheta \in \mathbb{R}\}$ be a family of distribution functions differing in location (i.e. $F(x,\vartheta) = F_o(x-\vartheta)$; such a family will be called a location model). Huber's *M-estimator* $T(F_n)$ is defined by the equation

$$\int \psi(x - T(F_n))\ dF_n(x) = 0$$

where ψ denotes some real valued measurable function. We shall make the following assumptions:

- ψ is differentiable
- For some neighbourhood U of 0 and for some integrable function u $\sup_{\vartheta \in U} |\psi'(x-\vartheta)| \leq u(x) \quad (x \in \mathbb{R})$
- $\int \psi(x)\ dF(x) = 0 \quad$ where $F = F_o$
- $\int \psi'(x)\ dF(x) \neq 0$.

Since

$$\frac{\partial}{\partial\vartheta} \int \psi(x-\vartheta)\ dF^{(t)}(x) = - \int \psi'(x-\vartheta)\ dF^{(t)}(x),$$

$T(F^{(t)})$ is well defined for all $0 \leq t \leq 1$ with a probability tending to 1 as $n \longrightarrow \infty$. Implicit differentiation shows that

$$\frac{d}{dt}T(F^{(t)})\Big|_{t=s} = \left(\int \psi'(x-T(F^{(s)}))\, dF^{(s)}(x) \right)^{-1}$$

$$\int \psi(x-T(F^{(s)}))\, d(F_n-F)(x)$$

and T is a differentiable statistical functional of order 1 in F. More details about M-estimators can be found in section 3 of this chapter.

Example 2.1.7: The *p-th quantile (fractile)* of a d.f. F is defined by

$$T(F) = F^{-1}(p) = \inf \{ x : F(x) \geq p \}.$$

Let us first assume that for some fixed d.f. F and G and for some $0 \leq t \leq 1$ (abusing the notation for a moment)

(1) $F_t = F + t(G-F)$ is differentiable in $F_t^{-1}(p)$ and $F_{1/2}$ is continuous in some neighbourhood of $F_t^{-1}(p)$.

(2) F_t^{-1} is continuous in p.

We claim first that $s \longrightarrow F_s^{-1}(p)$ is differentiable in $s = t$ and that

$$\text{(2.1.11)} \quad \frac{d}{ds}F_s^{-1}\Big|_{s=t} = \left\{ \frac{\partial F_t(x)}{\partial x}\Big|_{x=F_t^{-1}(p)} \right\}^{-1} (F(F_t^{-1}(p)) - G(F_t^{-1}(p))).$$

Observe that (2) implies the continuity of $s \longrightarrow F_s^{-1}(p)$ in $s = t$. Indeed, if $x = F_{t+h}^{-1}(p)$, then

$$p \leq F_{t+h}(x) = F_t(x) + h(G(x)-F(x)) .$$

The continuity of F_t^{-1} in p implies

$$F_t^{-1}(p) \leq \liminf_{h \longrightarrow 0} F_{t+h}^{-1}(p).$$

On the other hand, if $x < F_{t+h}^{-1}(p)$, then $p > F_t(x) + h(G(x) - F(x))$ and $x \leq F_t^{-1}(p+|h|)$. Letting $x \longrightarrow F_{t+h}^{-1}(p)$ it follows that

$$\limsup_{h \longrightarrow 0} F_{t+h}^{-1}(p) \leq F_t^{-1}(p) .$$

Observe next that $F_{t+h}(F_{t+h}^{-1}(p)) = p$ if h is sufficiently small. Otherwise F or G is not continuous in every neighbourhood of $F_t^{-1}(p)$, contradicting (1).

For proving (2.1.11) let $h \neq 0$ be sufficiently small. Then

$$\frac{F_{t+h}^{-1}(p) - F_t^{-1}(p)}{h} = \frac{F_t(F_{t+h}^{-1}(p)) - p}{h} \; \frac{F_{t+h}^{-1}(p) - F_t^{-1}(p)}{F_t(F_{t+h}^{-1}(p)) - p} =$$

$$= \left(F(F_{t+h}^{-1}(p)) - G(F_{t+h}^{-1}(p))\right)\left\{ \frac{F_t(F_{t+h}^{-1}(p)) - F_t(F_t^{-1}(p))}{F_{t+h}^{-1}(p) - F_t^{-1}(p)} \right\}^{-1}$$

$$\longrightarrow \left\{ \frac{\partial}{\partial x} F_t(x) \Big|_{x=F_t^{-1}(p)} \right\}^{-1} \left(F(F_t^{-1}(p)) - G(F_t^{-1}(p))\right)$$

since $F_{t+h}^{-1}(p) \longrightarrow F_t^{-1}(p)$ as $h \longrightarrow 0$ and since F and G are continuous at $F_t^{-1}(p)$.

Let us now assume that F is differentiable in $F^{-1}(p)$ and that F is continuous and strictly monotone in some neighbourhood of $F^{-1}(p)$. Then F^{-1} is continuous in p. Moreover, if $G = F_n$ denotes the empirical d.f. of n independent, F-distributed observations, then F_n is a.s. continuous at $F^{-1}(p)$. Given $\eta > 0$ there exists a neighbourhood U of $F^{-1}(p)$ such that with probability $> 1 - \eta$ F_n is continuous in U. In this case it is easily checked that (1) and (2) are satisfied (for any t). Especially,

$$\frac{d}{dt} T(F_t) \Big|_{t=0} = \frac{p - F_n(F^{-1}(p))}{F'(F^{-1}(p))} =$$

$$= - (F'(F^{-1}(p)))^{-1} \int 1_{(-\infty, F^{-1}(p)]}(x) \, d(F_n - F)(x)$$

and hence T is a differentiable statistical functional of order 1 at F.

Example 2.1.8: The α-*trimmed mean* $(0 < \alpha < 1/2)$ can be defined by

$$T(F) = (1 - 2\alpha)^{-1} \int_{\alpha}^{1-\alpha} F^{-1}(x) \, dx =$$

$$= (1-2\alpha)^{-1} \left[\int_{F^{-1}(\alpha)}^{F^{-1}(1-\alpha)} x \, dF(x) + F^{-1}(\alpha)(F(F^{-1}(\alpha)) - \alpha) + \right.$$

$$\left. + F^{-1}(1-\alpha)(1-\alpha-F(F^{-1}(1-\alpha))) \right]$$

where, as before, F^{-1} denotes the left continuous inverse of F. If F_n denotes the empirical d.f. of n independent, F-distributed random variables with continuous d.f. F, then

$$T(F_n) = \frac{1}{n(1-2\alpha)} \sum_{\alpha n<k\leq (1-\alpha)n} X_{(k)}$$

where $X_{(1)} < X_{(2)} < \ldots < X_{(n)}$ denotes the order statistic.
From this formula it is clear that the α-trimmed mean may be considered in between the sample mean $(\alpha \longrightarrow 0)$ and the sample median $(\alpha \longrightarrow 1/2)$.
Assume that F is strictly monotone and differentiable with bounded derivative f. Let $F^{(t)} = F + t\,(F_n - F)$. Then

$$h^{-1} \,|F^{(h)^{-1}}(x) - F^{-1}(x)| \leq \sup_y |f(y)| < \infty$$

and using the previous example it follows that

$$\frac{d}{dt}T(F^{(t)})\Big|_{t=0} = -\frac{1}{1-2\alpha} \int_{\alpha}^{1-\alpha} (f(F^{-1}(x))^{-1}$$

$$\int 1_{(-\infty,F^{-1}(x)]}(y)\; d(F_n-F)(y)\; dx\;.$$

Therefore, the α-trimmed mean is a differentiable statistical functional of order 1 in F.

Example 2.1.9: The U-statistics of chapter 1 may be considered as von Mises' functionals. Let $h : \mathbb{R}^m \longrightarrow \mathbb{R}$ be a symmetric kernel of degree m and let F be a continuous d.f. Assume w.l.o.g. that $h(x_1,\ldots,x_m) = 0$ if for some $1\leq i\neq j\leq m$ $x_i = x_j$. In this case we shall say that h vanishes on the diagonals. We write the U-statistic as a V-statistic in the form

$$n^{-m} \binom{n}{m} m!\; U_n(h) = \int\ldots\int h(x_1,\ldots,x_m) \prod_{i=1}^{m} dF_n(x_i).$$

The right-hand side is a special value of the functional T defined by

$$T(F) = \int\ldots\int h(x_1,\ldots,x_m) \prod_{i=1}^{m} dF(x_i).$$

For any two d.f. F and G and for $F^{(t)} = F + t\,(G - F)$ we have

$$T(F^{(t)}) = \sum_{k=0}^{m} \binom{m}{k} \int\ldots\int h(x_1,\ldots,x_m) \prod_{i=1}^{k} dF(x_i) \prod_{j=k+1}^{m} d(G-F)(x_j)\, t^{m-k}$$

and therefore

$$\frac{d^p}{dt^p} T(F^{(t)}) \Big|_{t=s} = \sum_{k=o}^{m-p} \binom{m}{k} (m-k)(m-k-1)\ldots(m-k-p+1)\, s^{m-k-p}$$

$$\int \ldots \int h(x_1,\ldots,x_m) \prod_{i=1}^{k} dF(x_i) \prod_{j=k+1}^{m} d(G-F)(x_j)$$

This is a von Mises' functional of order m. (exercise: Verify the details using the results in chapter 1.)

It is not necessary to assume that h vanishes on the diagonals for defining T. Thus V-statistics defined in Definition 1.1.3 also are von Mises' functionals.

2. The asymptotic distribution of differentiable statistical functionals

We start this section introducing multiple stochastic integrals with respect to the Brownian bridge. (These integrals also could be defined using multiple Wiener-Ito integrals, but this will turn out to be unnecessarily complicated.)

Let $m \in \mathbb{N}$ and let $f : \mathbb{R}^m \longrightarrow \mathbb{R}$ be a measurable function. For a partition $Q = \{Q_1,\ldots,Q_q\}$ of $\{ 1,\ldots,m \}$ define $f_Q : \mathbb{R}^q \longrightarrow \mathbb{R}$ by

$$f_Q(x_1,\ldots,x_q) = f(y_1,\ldots,y_m)$$

where $y_j = x_i$ if and only if $j \in Q_i \in Q$ $(1 \leq j \leq m,\ 1 \leq i \leq q)$. f_Q is a measurable function on $\mathbb{R}^q$. If F is a distribution function denote by $L_2(m,F)$ the set (of equivalence classes) of all measurable functions $f : \mathbb{R}^m \longrightarrow \mathbb{R}$ such that for every partition Q of $\{1,\ldots,m\}$ the function f_Q is square integrable with respect to the q-fold product measure F^q of F ($q = |Q|$). Denoting by $|f_Q|$ the $L_2(F^q)$-norm of f_Q the norm of f in $L_2(m,F)$ is defined by

$$\|f\| = \sum |f_Q| \tag{2.2.1}$$

where the summation extends over all partitions Q of $\{1,\ldots,m\}$. For example, if m = 2, then

$$\|f\| = \left(\int\int f^2(x,y)\ dF(x)\ dF(y) \right)^{1/2} +$$

$$+ \left(\int f^2(x,x)\ dF(x) \right)^{1/2} .$$

We have $L_2(m,F) = \{ f : \mathbb{R}^m \longrightarrow \mathbb{R} : \|f\| < \infty \}$.

Lemma 2.2.1: For every $m \in \mathbb{N}$ and for every d.f. F the set $L_2(m,F)$ is a Banach space with norm $\| \ \|$ defined in (2.2.1).

Proof: It is obvious that $(L_2(m,F), \| \ \|)$ is a normed vector space. We show completeness. Let $f_n \in L_2(m,F)$ be a Cauchy sequence. Then for every partition Q, $(f_n)_Q$ is Cauchy in $L_2(F^q)$ and therefore convergent to some $f_Q \in L_2(F^q)$. Define

$$f(x_1,\ldots,x_m) = f_Q(y_1,\ldots,y_q) \qquad (x \in \mathbb{R}^m)$$

where Q denotes the partition of $\{1,\ldots m\}$ determined by those indices of x for which the x-coordinates are equal and where y_j stands for the common value of the x_i with $i \in Q_j$. It is evident that f_n converges to f in the norm $\| \ \|$.

Lemma 2.2.2: The set of simple functions h of the form

$$(2.2.2) \qquad h = \sum_{i_1,\ldots,i_m=0}^{s} h_{i_1,\ldots,i_m} \prod_{j=1}^{m} 1_{(t_{i_j}, t_{i_j+1}]}$$

is dense in $L_2(m,F)$, where $h_{i_1,\ldots,i_m} \in \mathbb{R}$ and

$-\infty = t_0 < t_1 < \ldots < t_{s+1} = \infty$ is a partition of $\mathbb{R}$.

Proof: Let $f \in L_2(m,F)$. Assume first that f is bounded, continuous and has compact support. For $\varepsilon > 0$ choose a partition $(t_j)_{0 \leq j \leq s+1}$ such that f varies at most by ε on each of the rectangles $\prod (t_{i_j}, t_{i_j+1}]$. Let $h_{i_1,\ldots,i_m}$ be any value of f in this rectangle. Then the simple function h defined by (2.2.2) with the above choices satisfies

$\|f - h\| \leq \varepsilon \cdot$number of partitions of $\{1,\ldots,m\}$.

Let now f be bounded. By Lusin's theorem there exists a compact set $K \subset \mathbb{R}^m$ such that $f|_K$ is continuous and

$F_Q(K) > 1-\eta$ for all partitions Q where F_Q denotes the product measure F^q on the set

$\Delta_Q = \{ (x_1,\ldots,x_m) : x_i = x_j$ iff i and j belong to the same member of $Q \}$.

Here $\eta > 0$ is arbitrarily small. Using the first case it is easy to see that for $\varepsilon > 0$ there exists a simple function of the form (2.2.2) such that $\|f - h\| < \varepsilon$ provided η is small enough.

Finally, an arbitrary function $f \in L_2(m,F)$ can be approximated using the last case and a truncation argument for each product measure F^q.

Lemma 2.2.3: Let $h \in L_2(F^2)$ be a symmetric function. Then there exist orthonormal functions $f_j \in L_2(F)$ $(j \geq 1)$ and $\lambda_j \in \mathbb{R}$ satisfying $\Sigma\lambda_j^2 < \infty$ and

$$h(x,y) = \sum_{j\geq 1} \lambda_j f_j(x) f_j(y)$$

in $L_2(F^2)$.

Proof: The operator $A : L_2(F) \longrightarrow L_2(F)$ defined by

$$Af(x) = \int h(x,y) f(y) dF(y) \qquad (f \in L_2(F))$$

is a self-adjoint Hilbert-Schmidt operator. Denote by λ_j the eigenvalues of A and by f_j the corresponding normed, orthogonal eigenvectors $(j \geq 1)$. Then

$$Af = \sum_{j\geq 1} \lambda_j \langle f,f_j\rangle f_j$$

and for the Hilbert-Schmidt norm $|A| = \left(\sum_{j\geq 1} |Af_j|^2 \right)^{1/2}$

one has $\Sigma \lambda_j^2 = |A|^2 < \infty$. Moreover

$$\int\int \left(\sum_{j\geq 1} \lambda_j f_j(x) f_j(y)\right)^2 dF(x) dF(y) = \sum_{j\geq 1} \lambda_j^2 < \infty$$

and for $f \in L_2(F)$

$$\int \sum_{j\geq 1} \lambda_j f_j(x) f_j(y) f(y) dF(y) = \sum_{j\geq 1} \lambda_j \langle f_j,f\rangle f_j(x) =$$

$$= Af(x) = \int h(x,y) f(y) dF(y) .$$

Let h be a simple function as introduced in (2.2.2) considered as an element of $L_2(m,\lambda)$ (i.e. $t_o=0$, $t_{s+1}=1$) where λ denotes the Lebesgue measure on $[0,1]$. Denote by $\{B(t) : 0 \leq t \leq 1\}$ the Brownian bridge. Recall that B is a Gaussian process on $C([0,1])$ satisfying $EB(t) = 0$ $(0 \leq t \leq 1)$ and $EB(t)B(s) = s(1-t)$ $(0 \leq s \leq t \leq 1)$.

The stochastic integral of the simple function h with respect to B is defined by

$$(2.2.3) \qquad Z(h) = \int\cdots\int h(x_1,\ldots,x_m) \prod_{i=1}^{m} dB(x_i) =$$

$$= \sum_{i_1,\dots,i_m=0}^{s} h_{i_1,\dots,i_m} \prod_{j=1}^{m} (B(t_{i_j+1}) - B(t_{i_j})) .$$

This definition can be extended to $L_2(m,\lambda)$ according to

<u>Theorem 2.2.1:</u> The map $h \longrightarrow Z(h)$ defined in (2.2.3) has a unique extension to a bounded linear operator

$$Z : L_2(m,\lambda) \longrightarrow L_2(P)$$

where $(\Omega,\mathcal{F},P)$ denotes the probability space on which the Brownian bridge is defined. We also shall write

$$Z(h) = \int \dots \int h(x_1,\dots,x_m)\, dB(x_1)\dots dB(x_m) .$$

The purpose of Z is to describe weak limit distributions. Thus it would be sufficient to consider just one fixed probability space.

<u>Proof:</u> We have to show that Z(h) exists and that

$$(2.2.4) \qquad |Z(h)|_{L_2(P)} \leq C \, \|h\| \qquad (h \in L_2(m,\lambda))$$

where C denotes some constant only depending on m and where $|\ .\ |_{L_2(P)}$ denotes the norm in $L_2(P)$. This will be accomplished by showing (2.2.4) for simple functions of the form (2.2.2) in view of Lemma 2.2.2.

Let h be a simple function as in (2.2.2). Then

$$E\, Z(h)^2 = \sum_{i_1,\dots,i_{2m}=0}^{s} h_{i_1,\dots,i_m} h_{i_{m+1},\dots,i_{2m}} E \prod_{j=1}^{2m} \beta(i_j)$$

where $\beta(l) = B(t_{l+1}) - B(t_l)$ $(l = 0,\dots,s)$. The characteristic function φ of the vector $(\beta(0),\dots,\beta(s))$ is given by

$$(2.2.5) \qquad \varphi(u) = E \exp\left(i \sum_{k=0}^{s} u_k \beta(k)\right) = \exp(- \tfrac{1}{2} uSu')$$

where $u = (u_0,\dots,u_s) \in \mathbb{R}^{s+1}$ and where $S = (s_{ij})_{0 \leq i,j \leq s}$ denotes the covariance operator of the Gaussian random vector $(\beta(0),\dots,\beta(s))$. It is easily seen that for $i < j$

$$s_{ij} = E\beta(i)\beta(j) = - (t_{i+1} - t_i)\,(t_{j+1} - t_j)$$

and that for $i = j$

$$s_{ij} = (t_{i+1} - t_i) - (t_{i+1} - t_i)^2 .$$

Differentiating both sides in (2.2.5) we obtain

$$\left.\frac{\partial^{2m}\varphi}{\partial u_{i_1}\dots\partial u_{i_{2m}}}\right|_{u_0=\dots=u_s=0} = \left.\frac{\partial^{2m}\exp(-2^{-1}uSu')}{\partial u_{i_1}\dots\partial u_{i_{2m}}}\right|_{u_0=\dots=u_s=0}$$

and the left-hand side of this equality is $i^{2m}\, E \prod_{j=1}^{2m} \beta(i_j)$.
We claim that

$$(2.2.6)\qquad \frac{\partial^n \exp(-2^{-1}uSu')}{\partial u_{i_1}\dots\partial u_{i_n}} =$$

$$= \sum_{k=0}^{[n/2]} (-1)^{n-k} \sum\nolimits_k s_{i_{l_1} i_{l_2}} \dots s_{i_{l_{2k-1}} i_{l_{2k}}} S_{i_{l_{2k+1}}}(u) \dots S_{i_{l_n}}(u)\; f(u)$$

where (i) $f(u) = \exp(-1/2\ uSu')$,
(ii) $S_j(u) = \sum_{0 \leqq k \leqq s} s_{jk} u_k \qquad (0 \leqq j \leqq s)$ and
(iii) $\sum_k$ denotes the summation over all choices of pairs $(l_1,l_2),\dots,(l_{2k-1},l_{2k})$ and indices $l_{2k+1},\dots,l_n$ such that $\{l_1,\dots,l_n\} = \{1,\dots n\}$, $\quad l_i < l_{i+1}$ $(i=1,3,\dots,2k-1,2k+1,\dots n)$ and $l_i < l_{i+2}$ $(i=1,3,5,\dots,2k-1)$.

We shall prove (2.2.6) by induction over n.

$\frac{\partial f}{\partial u_i} = -\,S_i(u)\, f(u)$ shows (2.2.6) for $n = 1$. Assume now that (2.2.6) holds for n. Let $\delta(n) = 1$ if n is odd and $\delta(n) = 0$ if n is even. Then by induction hypothesis

$$\frac{\partial^{n+1} f}{\partial u_{i_1}\dots\partial u_{i_{n+1}}} = \sum_{k=0}^{[n/2]} (-1)^{n-k} \sum\nolimits_k s_{i_{l_1} i_{l_2}} \dots s_{i_{l_{2k-1}} i_{l_{2k}}}$$

$$\left(\sum_{j=2k+1}^{n} s_{i_{l_j} i_{n+1}} \prod_{r \neq j} S_{i_{l_r}}(u)\right) f(u) \quad -$$

$$- \sum\nolimits_k s_{i_{l_1} i_{l_2}} \dots s_{i_{l_{2k-1}} i_{l_{2k}}} \left(\prod_{j=2k+1}^{n} S_{i_{l_j}}(u)\, S_{i_{n+1}}(u)\right) f(u) =$$

$$= \sum_{k=1}^{[(n+1)/2]} (-1)^{n+1-k} \sum_{\substack{k \\ n+1 \in \{l_1,\dots,l_{2k}\}}} s_{i_{l_1} i_{l_2}} \dots s_{i_{l_{2k-1}} i_{l_{2k}}}$$

$$f(u) \prod_{j>2k} S_{i_{l_j}}(u) \; +$$

$$+ \sum_{k=o}^{[n/2]} (-1)^{n+1-k} \sum_{k\; n+1\notin\{l_1,\dots,l_{2k}\}} s_{i_{l_1} i_{l_2}} \cdots s_{i_{l_{2k-1}} i_{l_{2k}}}$$

$$f(u) \prod_{j>2k} S_{i_{l_j}}(u) =$$

$$= (-1)^{n+1} \prod_{j=1}^{n+1} S_{i_j}(u)\; f(u) + \delta(n)\; (-1)^{n+1-[(n+1)/2]}$$

$$f(u) \sum_{\left[\frac{n+1}{2}\right]} s_{i_{l_1} i_{l_2}} \cdots s_{i_{l_n} i_{l_{n+1}}}$$

$$+ \sum_{k=1}^{[n/2]} (-1)^{n+1-k} \sum_k s_{i_{l_1} i_{l_2}} \cdots s_{i_{l_{2k-1}} i_{l_{2k}}} \quad f(u) \prod_{j>2k} S_{i_{l_j}}(u).$$

(2.2.6) follows from this last expression.
Replacing in (2.2.6) n by 2m and evaluating the right-hand side for $u_0 = \dots = u_s = 0$ it follows that

$$i^{-2m} \left. \frac{\partial^{2m} f}{\partial u_{i_1} \cdots \partial u_{i_{2m}}} \right|_{u_0=\dots=u_s=0} =$$

$$= \sum_m s_{i_{l_1} i_{l_2}} \cdots s_{i_{l_{2m-1}} i_{l_{2m}}} = E \prod_{j=1}^{2m} \beta(i_j) .$$

Consequently we can write

$$E\; Z(h)^2 = \sum_m \sum_{i_1,\dots,i_{2m}=o}^{s} h_{i_1 \dots i_m} h_{i_{m+1} \dots i_{2m}}$$

$$s_{i_{l_1} i_{l_2}} \cdots s_{i_{l_{2m-1}} i_{l_{2m}}} .$$

Here the inner sum is bounded by

$$\int \cdots \int |h(x_1,\dots,x_m) h(x_{m+1},\dots,x_{2m})|$$

$$dp(x_{l_1}, x_{l_2}) \cdots dp(x_{l_{2m-1}}, x_{l_{2m}})$$

where $dp(x,y) = dxdy + \delta(x,y)\, dx$ with $\delta(x,y) = 1$ if $x = y$ and 0 otherwise. Using

$$|h(x_1,\dots,x_m) h(x_{m+1},\dots,x_{2m})| \leq 1/2(h^2(x_1,\dots,x_m) +$$

$$+ h^2(x_{m+1},\dots,x_{2m}))$$

and the definition of $\|h\|$, it is now easy to show (2.2.4) for simple functions of the form (2.2.2) . This proves the theorem.

Proposition 2.2.1:

(1) Let $h \in L_2(1,\lambda)$. Then the distribution of $Z(h)$ is normal with expectation zero and variance

$$\sigma^2 = \int h^2(x)\, dx - \left(\int h(x)\, dx \right)^2 .$$

(2) Let $h \in L_2(2,\lambda)$ be symmetric and degenerate. Then

$$(2.2.7) \quad Z(h) = \sum_{j\geq 1} \lambda_j \left(\left\{ \int f_j(x)\, dB(x) \right\}^2 - 1 \right) + $$
$$+ \int\int 1_{\{x=y\}}\, h(x,y)\, dB(x)\, dB(y) \quad \text{a.s.}$$

where $h(x,y) = \sum_{j\geq 1} \lambda_j f_j(x) f_j(y)$ denotes the representation of h in $L_2(\lambda^2)$ derived in Lemma 2.2.3. The random variables $\{ \int f_j\, dB : j \geq 1\}$ are independent, standard normal variables.

Proof: (1) If h is a simple function as in (2.2.2) then $Z(h) = \sum_{i=0}^{s} h_i\, (B(t_{i+1}) - B(t_i))$. Hence $Z(h)$ is a Gaussian random variable with expectation zero and variance

$$\sigma^2 = E\, Z(h)^2 = \sum_{i,j=0}^{s} h_i h_j\, E(B(t_{i+1})-B(t_i))(B(t_{j+1})-B(t_j)) =$$
$$= \sum_{i=0}^{s} h_i^2\, (t_{i+1}-t_i) - \sum_{i,j=0}^{s} h_i h_j\, (t_{i+1}-t_i)(t_{j+1}-t_j) =$$
$$= \int h^2(x)\, dx - \left(\int h(x)\, dx \right)^2.$$

In view of Lemma 2.2.2 and Theorem 2.2.1 this is sufficient to prove (1).

(2) Since $h \in L_2(2,\lambda)$ it follows in particular that $h \in L_2(\lambda^2)$ and by Lemma 2.2.3 $h(x,y) = \sum \lambda_j f_j(x) f_j(y)$ in $L_2(\lambda^2)$. By (1) and by the degeneracy and orthonormality of the functions f_j the random variables $\int f_j\, dB$ are independent and standard normal. In fact, if $i \neq j$, then

approximating f_j and f_i by simple functions h_j and h_i (with respect to the same partition) we have

$$E \int f_j \, dB \int f_i \, dB = \lim - \sum_{k,l=0}^{s} h_{jk} h_{il} (t_{k+1} - t_k)(t_{l+1} - t_l) + $$

$$+ \lim \sum_{k=0}^{s} h_{jk} h_{ik} (t_{k+1} - t_k) =$$

$$= \int f_j(x) f_i(x) \, dx - \int f_j(x) \, dx \int f_i(x) \, dx = 0.$$

(Observe the following: Since f_j is an eigenfunction of the Hilbert-Schmidt operator w.r.to λ_j one has by degeneracy that

$$\int \lambda_j f_j(x) dx = \iint h(x,y) f_j(y) dy \, dx = 0 \,.)$$

Observe that $\sum_{j \geq 1} \lambda_j ((\int f_j \, dB)^2 - 1)$ is well defined since

$$E \left(\sum_{j \geq 1} \lambda_j \left[\left\{ \int f_j \, dB \right\}^2 - 1 \right] \right)^2 =$$

$$= \sum_{j \geq 1} \lambda_j^2 \, E \left((\int f_j \, dB)^2 - 1 \right)^2 < \infty \quad .$$

Let us now prove (2) by approximation of f_j with simple functions

$$g_j = \sum a_{kj} 1_{(t_k, t_{k+1}]} \qquad (j \geq 1)$$

and of $x \longrightarrow h(x,x)$ with simple functions

$$g = \sum a_k 1_{(t_k, t_{k+1}]} \quad .$$

Define

$$\bar{g}_j(x,y) = \begin{cases} 0 & \text{if for some } k \;\; t_k < x,y \leq t_{k+1} \\ g_j(x) \, g_j(y) & \text{otherwise} \end{cases}$$

and

$$\psi(x,y) = \sum_{j \leq K} \lambda_j \, \bar{g}(x,y) + \sum a_k 1_{(t_k, t_{k+1}] \times (t_k, t_{k+1}]}(x,y).$$

Note that $\int (\psi(x,x) - h(x,x))^2 \, dx \longrightarrow 0$. Since also $\sum a_k^2 (t_{k+1} - t_k)^2 \longrightarrow 0$, it follows that $\| \sum \lambda_j \bar{g}_j - h \| \longrightarrow 0$ in $L_2(\lambda^2)$ and therefore $\| \psi - h \| \longrightarrow 0$ in $L_2(2,\lambda)$ as $K \longrightarrow \infty$ and g, g_j approach their limits.

By Theorem 2.2.1 $Z(\psi) \longrightarrow Z(h)$ in $L_2(W)$.

Therefore it is left to show that $Z(\psi)$ also approximates the right-hand side of (2.2.7) in $L_2(W)$. Observe that

$$(2.2.8) \quad E\Big(Z(\psi) - \sum_{j\leq K} \lambda_j \Big\{ \Big(\int f_j \, dB\Big)^2 - 1\Big\} - \\ - \iint 1_{\{x=y\}} \, h(x,y) \, dB(x) \, dB(y) \Big)^2 \leqq$$

$$\leqq 2 \, E\Big(\iint 1_{\{x=y\}} \, (g(x)-h(x,y)) \, dB(x) \, dB(y) \Big)^2 +$$

$$+ 2 \, E\Big(\sum_{j\leq K} \lambda_j \Big\{ \iint \bar{g}_j(x,y) \, dB(x) dB(y) - \Big(\int f_j dB\Big)^2 + 1 \Big\} \Big)^2 \leq$$

$$\leq 2C \, \|h-g\|^2 + 2 \, E\Big(\sum_{j\leq K} \lambda_j \Big\{ \Big(\int g_j \, dB\Big)^2 - \Big(\int f_j \, dB\Big)^2 + 1 - \\ - \sum a_{kj}^2 (B(t_{k+1})-B(t_k))^2 \Big\} \Big)^2 .$$

Since $g_j \longrightarrow f_j$ in $L_2(\lambda)$ and since $\int g_j \, dB$ and $\int f_j \, dB$ are normal, we have

$$\Big(\int g_j \, dB\Big)^2 - \Big(\int f_j \, dB\Big)^2 \longrightarrow 0 \quad (\text{in } L_2(W)).$$

Moreover,

$$E\Big(1 - \sum a_{kj}^2 \, (B(t_{k+1})-B(t_k))^2\Big)^2 =$$

$$= 1 - 2 \sum a_{kj}^2 \, (t_{k+1}-t_k) + \sum a_{kj}^2 a_{lj}^2 (t_{k+1}-t_k) \\ (t_{l+1}-t_l) + o(1) =$$

$$= \Big(1 - \int g_j^2(x) \, dx\Big)^2 + o(1) \longrightarrow 0.$$

(Exercise: Check the details using the formulas in the proof of Theorem 2.2.1.)

Letting $K \longrightarrow \infty$, $g_j \longrightarrow f_j$ and $g(x) \longrightarrow h(x,x)$ it follows that (2.2.8) tends to zero. This completes the proof of (2.2.7).

A similar inequality as (2.2.4) holds for stochastic integrals with respect to the empirical process, a result to be proven below.

Let $h \in L_2(m,F)$ and let F_n denote the empirical d.f. of n independent, F-distributed random variables $X_1,\ldots,X_n$. Recall that the stochastic integral of h with respect to F_n was defined as a V-statistic in Definition 1.1.3:

$$V_n = V_n(h) = \int \dots \int h(x_1,\dots,x_m) \prod_{i=1}^{m} dF_n(x_i) =$$

$$= n^{-m} \sum_{i_1,\dots,i_m=1}^{n} h(X_{i_1},\dots,X_{i_m}) .$$

The function h was called degenerate if

$$\int h(x_1,\dots,x_m)\, dF(x_i) = 0$$

for all $i = 1,\dots,m$ and all x_j $(1 \leq j \neq i \leq m)$. It follows from the results in chapter 1, section 2, that stochastic integrals

$$\int \dots \int h(x_1,\dots,x_m) \prod_{i=1}^{m} d(F_n - F)(x_i)$$

can be written as V-statistics $V_n(\tilde{h})$ setting

$$\tilde{h} = \sum_{K \subset \{1,\dots,m\}} (-1)^{|K|}\, h_K \circ \mathrm{proj}_K$$

where

$$h_K(y_1,\dots,y_{|K|}) = \int \dots \int h(x_1,\dots,x_m) \prod_{i \notin K} dF(x_i) \text{ with } x_i = y_i\ (i \in K).$$

The inequality similar to (2.2.4) is contained in

Proposition 2.2.2: For every $m \in \mathbb{N}$ there exists a constant C such that: If $h \in L_2(m,F)$ is degenerate , where F denotes a d.f., then

$$\sup_{n \in \mathbb{N}} E\left(n^{m/2}\, V_n(h) \right)^2 \leq C\, \|h\|^2. \tag{2.2.9}$$

Proof: Write $n^{2m}\, E(V_n(h))^2$ in the form

$$\sum_{i_1,\dots,i_m=1}^{n} \ \sum_{j_1,\dots,j_m=1}^{n} Eh(X_{i_1},\dots,X_{i_m})\, h(X_{j_1},\dots,X_{j_m}) . \tag{2.2.10}$$

For a moment, let $i_1,\dots,i_m,j_1,\dots,j_m$ be fixed. If one index i_l is different from all the other indices i_k and also different from all $j_1,\dots,j_m$, then

$$E\, h(X_{i_1},\dots,X_{i_m})\, h(X_{j_1},\dots,X_{j_m}) =$$

$$= E\left(\int h(X_{i_1},\dots,X_{i_{l-1}},x,X_{i_l},\dots,X_{i_m}) h(X_{j_1},\dots,X_{j_m}) \, dF(x)\right)$$

$$= 0$$

by degeneracy. Similarly, the expectation is zero if one index j_l is different from all the others.
In general, applying Cauchy-Schwarz' inequality, we have

$$(E\, h(X_{i_1},\dots,X_{i_m})\, h(X_{j_1},\dots,X_{j_m}))^2 \leq$$

$$\leq E\, h^2(X_{i_1},\dots,X_{i_m}) \quad E\, h^2(X_{j_1},\dots,X_{j_m}) \;.$$

By the definition of the norm in $L_2(m,F)$ this last term is bounded by $\|h\|^4$, no matter which indices are equal.
Returning to (2.2.10) observe that the number of terms $Eh(X_{i_1},\dots,X_{i_m})h(X_{j_1},\dots,X_{j_m})$ which do not vanish is bounded by const. n^m. Consequently, (2.2.10) is bounded by $Cn^m\|h\|^2$ and therefore

$$n^m\, EV_n(h)^2 \leq C\, \|h\|^2.$$

Note that the constant is independent of n. (2.2.9) follows.
Remark: In order to prove (2.2.4) one could proceed differently as in the proof of Theorem 2.2.1. Using Proposition 2.2.2 , the weak convergence of $n^{m/2}\, V_n(h)$ to $Z(h)$ proven below for simple functions $h \in L_2(m,\lambda)$ and the uniform integrability of $n^m\, V_n(h)^2$ $(n \in \mathbb{N})$ (which is not too hard to verify), it follows that

$$E\, Z(h)^2 = \lim_{n\to\infty} n^m\, EV_n(h)^2 \leq C\, \|h\|^2$$

for simple functions h.
The results on the asymptotic behaviour of U-statistics, V-statistics and differentiable statistical functionals are easily established after the foregoing preparations. We begin with V-statistics.
Theorem 2.2.2: Let F be a d.f. and let $h \in L_2(m,F)$ be symmetric. Define

$$(2.2.11) \qquad \psi(x_1,\dots,x_m) = h(F^{-1}(x_1),\dots,F^{-1}(x_m))$$

for $x = (x_1,\dots,x_m) \in \mathbb{R}^m$. Then $\psi \in L_2(m,\lambda)$ and the distributions of

$$(2.2.12) \qquad n^{m/2} \int \dots \int h(x_1,\dots,x_m) \prod_{i=1}^{m} d(F_n - F)(x_i) \qquad (n \geq 1)$$

and

$$(2.2.13) \qquad n^{m/2} \int \dots \int \psi(x_1,\dots,x_m) \prod_{i=1}^{m} d(\lambda_n - \lambda)(x_i) \qquad (n \geq 1)$$

coincide, where F_n (resp. λ_n) denotes the empirical d.f. of n independent F-distributed (resp. λ-distributed) random variables.

Moreover, the random variables in (2.2.12) converge weakly to $Z(\psi)$ defined in Theorem 2.2.1.

Proof: The first part follows from a change of variables. We shall show the weak convergence of the random sequence in (2.2.13) to $Z(\psi)$. Let ψ be a simple function as in (2.2.2) and define a function $u : D([0,1]) \longrightarrow \mathbb{R}$ by

$$u(g) = \sum_{i_1,\dots,i_m=0}^{s} h_{i_1 \dots i_m} \prod_{j=1}^{m} (g(t_{i_j+1}) - g(t_{i_j})).$$

(Recall that $D([0,1])$ was used in Propositions 1.3.2 and 2.1.2) Since u is continuous at every continuous function $g \in D([0,1])$ and since $n^{1/2} (\lambda_n - \lambda)$ converges weakly to the Brownian bridge (which has continuous paths), it follows that $u(n^{1/2} (\lambda_n - \lambda))$ converges weakly to $u(B)$. Note that $u(n^{1/2} (\lambda_n - \lambda))$ is exactly (2.2.13) and that $u(B) = Z(\psi)$.

We have shown the weak convergence for simple functions and we shall extend the statement to arbitrary $\psi \in L_2(m,\lambda)$ using Theorem 2.2.1 and Proposition 2.2.2.

Approximate ψ by simple functions ψ_o. Using the linearity of the stochastic integrals (2.2.13) we obtain from Proposition 2.2.2 that

$$\sup_{n \in N} E \Big(n^{m/2} \int \psi(x_1,\dots,x_m) \prod_{i=1}^{m} d(\lambda_n - \lambda)(x_i) - $$

$$- n^{m/2} \int \psi_o(x_1,\dots,x_m) \prod_{i=1}^{m} d(\lambda_n - \lambda)(x_i) \Big)^2$$

$$\leq C \, \| \psi - \psi_o \|^2 .$$

Similarly, Theorem 2.2.1 (inequality (2.2.4)) implies that

$$E(Z(\psi) - Z(\psi_o))^2 \leq C \, \| \psi - \psi_o \|^2 .$$

The weak convergence of (2.2.13) to $Z(\psi)$ follows now from Lemma 2.2.2 and Chebychev's inequality.

<u>Theorem 2.2.3:</u> Let T be a differentiable statistical functional of order m at F. If $\varphi^{(m)} \in L_2(m,F)$, if

$$(2.2.14) \qquad \frac{d^p}{dt^p} T(F^{(t)}) \Big|_{t=0} \equiv 0$$

for $p = 1,\ldots,m-1$, and if

$$(2.2.15) \qquad n^{m/2} \left\{ T(F_n) - T(F) - \frac{1}{m!} \int \cdots \int \varphi^{(m)}(x_1,\ldots,x_m) \prod_{i=1}^{m} d(F_n - F)(x_i) \right\}$$

converges to zero in probability, then

$$n^{m/2} \; (T(F_n) - T(F)) \longrightarrow \frac{1}{m!} Z(\psi)$$

weakly as $n \longrightarrow \infty$ where ψ is defined by (2.2.11) with $h = \varphi^{(m)}$. The proof of this theorem is obvious in view of Theorem 2.2.2. Also using Definition 2.1.3 and Propositions 2.1.1 and 2.1.2 we have the following corollaries.

<u>Corollary 2.2.1:</u> Let T be a von Mises' functional of order m at F. Assume that (2.2.14) holds for $p = 1,\ldots,m-1$ and that $\varphi^{(m)} \in L_2(m,F)$. Then

$$n^{m/2} \; (T(F_n) - T(F)) \longrightarrow \frac{1}{m!} Z(\psi)$$

weakly where ψ is again defined by (2.2.11).

<u>Proof:</u> In order to apply Theorem 2.2.3 we have to verify that (2.2.15) converges to zero in probability. This follows immediately from (2.1.6) since

$$P\left(\left| n^{m/2} \left\{ T(F_n) - T(F) - \frac{1}{m!} \int \cdots \int \varphi^{(m)} \prod d(F_n - F) \right\} \right| \geq \varepsilon \right)$$

$$\leq P\left(n^{m/2} \sup_{0 \leq t \leq 1} \frac{1}{(m+1)!} \left| \frac{d^{m+1}}{ds^{m+1}} T(F^{(s)}) \Big|_{s=t} \right| \geq \varepsilon \right) \longrightarrow 0.$$

<u>Corollary 2.2.2:</u> Let the functional T be Fréchet differentiable at the continuous d.f. F. Define the function φ by $\varphi(x) = T'(\varepsilon_x)$ where ε_x denotes the point mass in $x \in \mathbb{R}$. If $\varphi \in L_2(F)$ then

$$n^{1/2} \; (T(F_n) - T(F)) \longrightarrow \mathcal{N}(0,\sigma^2)$$

weakly where

$$\sigma^2 = \int \varphi^2(x)\, dF(x) - \left(\int \varphi(x)\, dF(x) \right)^2.$$

Proof: It suffices to show that $T'(F_n - F) = \int \varphi(x)\, d(F_n - F)(x)$. Since T' is linear we have

$$T'(F_n) = n^{-1} \sum_{i \le n} T'(\varepsilon_{X_i}) = n^{-1} \sum_{i \le n} \varphi(X_i)$$

and

$$T'(F) = T'\left(\int \varepsilon_x\, dF(x) \right) = \int T'(\varepsilon_x)\, dF(x)$$

$$= \int \varphi(x)\, dF(x),$$

where $\int \varepsilon_x\, dF(x)$ is understood as a Bochner integral.

Corollary 2.2.3: Let T be a functional defined on an open subset U of $D([0,1])$ which is Hadamard differentiable at the uniform distribution $\lambda \in U$. Define φ as in Corollary 2.2.2. If $\varphi \in L_2(\lambda)$ then

$$n^{1/2} \left(T(\lambda_n) - T(\lambda) \right) \longrightarrow N(0, \sigma^2)$$

where $\sigma^2 = \int \varphi^2(x)\, dx - \left(\int \varphi(x)\, dx \right)^2.$

If T is a differentiable statistical functional of order $m \geq 1$ at F, the function $\varphi^{(1)}$ in Definition 2.1.1, (2.1.2), is called the *influence curve (or function)* of T at F. It is denoted by $IC(x,F,T)$. It determines the asymptotic variance in the preceding theorems (if $m=1$) and corollaries. It is not uniquely determined since $\varphi^{(1)}$ is not (only up to addition of constants) but from (2.1.2) it is clear that

$$IC(x,F,T) = \frac{d}{dt} T(F + t(\varepsilon_x - F)) \Big|_{t=0}$$

defines the influence curve unambiguously.

In the remaining part of this section we discuss the application of the preceding results to the examples of section 1.

Example 2.2.1: (Example 2.1.1 continued)

The sample central moment

$$T(F) = \int_{-\infty}^{\infty} (x - \mu_F)^k\, dF(x)$$

is a von Mises' functional of order 1 and

$$\frac{d}{dt}T(F^{(t)})\Big|_{t=0} = \int \left\{(x-\mu_F)^k - kx\int(z-\mu_F)^{k-1}dF(z)\right\} d(F_n-F)(x).$$

The influence curve is given by

$$IC(x,F,T) = (x-\mu_F)^k - kx \int (z-\mu_F)^{k-1} dF(z).$$

$n^{1/2}(T(F_n) - T(F))$ is asymptotically normal with expectation zero and variance

$$\sigma^2 = M_{2k} - M_k^2 - 2kM_{k-1}M_{k+1} + k^2M_{k-1}^2M_2$$

where

$$M_j = \int (F^{-1}(x) - \mu_F)^j dx \qquad (j \geq 1).$$

Example 2.2.2: (Example 2.1.2 continued)

Let $T(F)$ denote the maximum likelihood estimator. Under the assumptions in Example 2.1.2 it is a differentiable statistical functional. In order to apply Corollary 2.2.1 it is necessary to verify (2.1.6) for $p = 2$. This can be done under additional assumptions similarly to the case $p = 1$. However, it is easier to apply Theorem 2.2.3 directly since (2.2.15) can be verified quite easily. (In many examples a direct application of this theorem is preferable.)

Since

$$\frac{d}{dt}T(F^{(t)})\Big|_{t=0} = -\frac{1}{b(0)} \int g(x,\vartheta_o)\, d(F_n-F)(x)$$

it suffices to show that

$$n^{1/2} \sup_{0\leq t\leq 1} \left|\frac{d}{ds}T(F^{(s)})\Big|_{s=t} + \frac{1}{b(0)} \int g(x,\vartheta_o)\, d(F_n-F)(x)\right| \to 0$$

in probability.

It has been shown in Example 2.1.2 that $\sup b(t) \longrightarrow b(0) \neq 0$ in probability. What we have to show reduces moreover to

$$(2.2.16) \quad \sup_t n^{1/2} \left| \int (g(x,T(F^{(t)})) - g(x,\vartheta_o))\, d(F_n-F)(x)\right| \to 0$$

in probability.

Using Taylor's formula the integrand can be written as

$$tg'(x,\vartheta_o)\, \frac{d}{ds}T(F^{(s)})\Big|_{s=r} + \frac{t^2}{2}\, g''(x,\vartheta) \left(\frac{d}{ds}T(F^{(s)})\Big|_{s=r}\right)^2$$

where $0 \leq r \leq 1$ and where ϑ belongs to the interval with endpoints ϑ_o and $T(F^{(t)})$. Note that in Example 2.1.2 it has

also been shown that $n^{1/2} \sup \left| \frac{d}{ds} T(F^{(s)}) \Big|_{s=r} \right|$ is stochastically bounded. Since $|g''(x,\vartheta)| \leq w(x)$ with large probability (if n is sufficiently large) (2.2.16) follows.
Now Theorem 2.2.3 applies: $n^{1/2}\ (T(F_n)-T(F))$ is asymptotically normal $N(0,\sigma^2)$ with

$$\sigma^2 = \left\{ \int g'(x,\vartheta_o)\ dF(x) \right\}^{-2} \left(\int g^2(x,\vartheta_o)\ dF(x) - \right.$$

$$\left. - \left\{ \int g(x,\vartheta_o)\ dF(x) \right\}^2 \right) .$$

The influence curve is just

$$IC(x,F,T) = -\left(\int g'(x,\vartheta_o)\ dF(x) \right)^{-1} g(x,\vartheta_o).$$

<u>Example 2.2.3:</u> (Example 2.1.3 continued)
For the χ^2-minimum estimator one can proceed as in the previous example showing asymptotic normality. Since

$$\frac{d}{dt} T(F^{(t)}) \Big|_{t=0} = 2 \left(\iint \frac{\partial h(\vartheta,x,y)}{\partial\ \vartheta} \Big|_{\vartheta=\alpha} dF(x)\ dF(y) \right)^{-1}$$

$$\iint h(\alpha,x,y)\ dF(x)\ d(F_n-F)(y)$$

$n^{1/2}(T(F_n)-T(F))$ is asymptotically $N(0,\sigma^2)$ with

$$\sigma^2 = 4 \left\{ \int \left(\frac{\partial \log p(x,\vartheta)}{\partial\vartheta} \Big|_{\vartheta=\alpha} \right)^2 dF(x) \right\}^{-1}$$

where $p(x,\vartheta) = p_j(\vartheta)$ if $a_j < x \leq a_{j+1}$. The influence curve is defined by

$$IC(x,F,T) = \sigma^2/2 \int h(\alpha,x,y)\ dy.$$

<u>Example 2.2.4:</u> (Example 2.1.4 continued)
The χ^2-statistic χ^2 is a von Mises' functional T of any order at F with $\frac{d}{dt} T(F^{(t)}) \Big|_{t=0} = 0$. Therefore the limit distribution of $n\chi^2$ is given by the distribution of $1/2\ Z(\psi)$ where ψ denotes the kernel in $L_2(2,\lambda)$ corresponding to 2h. h was defined by

$$h(x,y) = \begin{cases} p_j^{-1}(\alpha) & \text{if } a_j < x,y \leq a_{j+1} \\ 0 & \text{otherwise} \end{cases} .$$

Thus the limit distribution is determined by

$$\iint h(F^{-1}(x),F^{-1}(y))\, dB(x)\, dB(y) =$$

$$= \sum_{j=1}^{r} \iint 1_{(a_j,a_{j+1}]}(F^{-1}(x))\, 1_{(a_j,a_{j+1}]}(F^{-1}(y))\, p_j^{-1}(\alpha)\, dB(x)dB(y) =$$

$$= \sum_{j=1}^{r} p_j^{-1}(\alpha)\, (B(F(a_{j+1})) - B(F(a_j)))^2 .$$

This representation does not yet show that the distribution of $T(F)$ is a χ^2-distribution with $r-1$ degrees of freedom, since the normal variables $B(F(a_{j+1})) - B(F(a_j))$ are not independent. This follows from the fact that h is not degenerate. But $h_o(x,y) = h(x,y) - 1$ is degenerate since

$$\int h_o(x,y)\, dF(y) = \sum_{j=1}^{r} p_j^{-1}(\alpha) \int 1_{(a_j,a_{j+1}]}(y)\, dF(y)$$

$$1_{(a_j,a_{j+1}]}(x) - 1 = 0.$$

We also have $Z(\psi) = Z(\psi_o)$ where ψ_o denotes the kernel in $L_2(2,\lambda)$ corresponding to $2h_o$. Set $h_o' = 1/2\ \psi_o$.
Thus we can proceed as in Proposition 2.2.1 (2). The Hilbert-Schmidt operator is given by

$$Af(x) = \int h_o'(x,y)\, f(y)\, dy =$$

$$= \sum_{j=1}^{r} p_j^{-1}(\alpha) \left[\int_{A_j} f(y)\, dy \right] 1_{A_j}(x) - \int f(y)\, dy =$$

$$= E\left(f - \int f(y)\, dy \;\middle|\; \mathcal{A}\right)$$

where $A_j = (F(a_j), F(a_{j+1})]$ and $\mathcal{A} = \sigma(A_1,\dots,A_r)$. Therefore A has an r-dimensional range. Since the dimension of all vectors $(c_1,\dots,c_r)$ with $\sum c_j\, p_j(\alpha) = 0$ is $r-1$ -dimensional and since for such a vector

$$A\left(\sum_{j=1}^{r} c_j\, 1_{A_j} \right) = \sum_{j=1}^{r} c_j\, 1_{A_j}$$

the eigenspace of the eigenvalue 1 is $r-1$-dimensional. Because of $A1 = 0$ it follows that

$$h_o'(x,y) = \sum_{j=1}^{r-1} f_j(x)\, f_j(y) \qquad \text{in } L_2(\lambda^2),$$

where the f_j are orthonormal eigenvectors of the eigenvalue 1.

It follows from Proposition 2.2.1 that

$$Z(h_o') = \sum_{j=1}^{r-1} \left(\int f_j \, dB \right)^2 - r + 1 +$$

$$+ \int\int 1_{\{x=y\}} h_o'(x,y) \, dB(x) \, dB(y) .$$

In order to show that $Z(h_o')$ has a χ^2-distribution with r-1 degrees of freedom it is left to show that

$$\int\int 1_{\{x=y\}} h_o'(x,y) \, dB(x) \, dB(y) = r-1 \quad \text{a.s.}$$

Partition $A_j = (F(a_j), F(a_{j+1})]$ into n intervals I_{jl} $(l = 1,\dots,n)$ of length $p_j(\alpha) \, n^{-1}$ and define

$$g_n(x,y) = \sum_{j=1}^{r} \sum_{l=1}^{n} (p_j^{-1}(\alpha) - 1) \, 1_{I_{jl}}(x) \, 1_{I_{jl}}(y) .$$

Then $\| g_n - 1_{\{x=y\}} h_o' \| \longrightarrow 0$ as $n \longrightarrow \infty$ and hence by Theorem 2.2.1 $Z(g_n) \longrightarrow Z(1_{\{x=y\}} h_o')$ in $L_2(W)$.

Now

$$Z(g_n) = \sum_{j=1}^{r} \sum_{l=1}^{n} (p_j^{-1}(\alpha) - 1) \, B(I_{jl})^2$$

where $B(I) = B(b) - B(a)$ for an interval $I = (a,b]$. Recall that for intervals $I = (a,b]$ and $J = (c,d]$

$$E \, B(I)^2 = (b - a)(1 - b + a)$$

and

$$E(B(I)^2 - EB(I)^2)(B(J)^2 - EB(J)^2) =$$

(2.2.17)
$$= \begin{cases} 2(b - a)^2 (d - c)^2 & \text{if } I \cap J = \emptyset \\ 2(b - a)^2 (1 - b + a)^2 & \text{if } I = J. \end{cases}$$

Hence

$$E \, Z(g_n) = \sum_{j=1}^{r} \sum_{l=1}^{n} (p_j^{-1}(\alpha) - 1) \, n^{-1} p_j(\alpha) \, (1 - n^{-1} p_j(\alpha)) =$$

$$= \sum_{j=1}^{r} (1 - p_j(\alpha))(1 - n^{-1} p_j(\alpha)) = r-1 + O(n^{-1}) \longrightarrow r-1$$

and $E(Z(g_n) - (r-1))^2 \longrightarrow 0$ will follow from

(2.2.18)
$$\lim_{n\to\infty} E(Z(g_n) - EZ(g_n))^2 = 0.$$

For fixed $j \in \{1,\dots,r\}$ we obtain using (2.2.17)

$$E \left\{ \sum_{l=1}^{n} (B(I_{jl})^2 - EB(I_{jl})^2) \right\}^2 =$$

$$= 2 \sum_{k \neq l} \frac{p_j^4(\alpha)}{n^4} + 2 \sum_{l=1}^{n} \frac{p_j^2(\alpha)}{n^2} (1 - n^{-1} p_j(\alpha))^2 \longrightarrow 0$$

as $n \longrightarrow \infty$ and this suffices to show (2.2.18).

Example 2.2.5: (Example 2.1.5 continued)

Let T(F) denote the ω^2-minimum estimator. We find that

$$\frac{d}{dt} T(F^{(t)}) \Big|_{t=0} = \left(\int \left\{ \frac{\partial F(x,\vartheta)}{\partial \vartheta} \Big|_{\vartheta=\vartheta_o} \right\}^2 f(x,\vartheta_o)\, dx \right)^{-1}$$

$$\int h(x)\, d(F_n - F)(x)$$

where $h(x) = \int \frac{\partial F(y,\vartheta)}{\partial \vartheta} \Big|_{\vartheta=\vartheta_o} 1_{(-\infty,x]}(y)\, f(y,\vartheta_o)\, dy.$

Therefore $n^{1/2}\,(T(F_n) - T(F))$ is asymptotically normal $N(0,\sigma^2)$ with

$$\sigma^2 = \left(\int \left\{ \frac{\partial F(x,\vartheta)}{\partial \vartheta} \Big|_{\vartheta=\vartheta_o} \right\}^2 f(x,\vartheta_o)\, dx \right)^{-1} \mathrm{Var}_F h.$$

If we consider the generalized Cramér-von Mises statistic

$$T(G) = \int w(F)\, (G - F)^2\, dF$$

we can apply Corollary 2.2.1 and find that the asymptotic distribution of $n\,(T(F_n) - T(F))$ is given by the random variable

$$2 \iint 1_{\{u<v\}} \int 1_{(F^{-1}(v),\infty)}(x)\, w(F(x))\, dF(x)\, dB(u)\, dB(v)$$

$$= \iint h(x,y)\, dB(x)\, dB(y)$$

with $h(x,y) = \int_{\{u > \max(x,y)\}} w(u)\, du$.

Example 2.2.6: (Example 2.1.6 continued)

Huber's M-estimator was defined by the equation

$$\int \psi(x - T(F_n))\, dF_n(x) = 0.$$

In general T is not a von Mises' functional, but under the assumptions in Example 2.1.6 it is a differentiable statistical functional of order 1 at F. We have seen already that

$$\frac{d}{dt}T(F^{(t)})\Big|_{t=s} = \left(\int \psi'(x-T(F^{(s)}))\, dF^{(s)}(x) \right)^{-1} \int \psi(x-T(F^{(s)}))\, d(F_n-F)(x).$$

Hence $n^{1/2}(T(F_n) - T(F))$ will be asymptotically normal with variance

$$\sigma^2 = \left(\int \psi'(x-\vartheta_o)\, dF(x) \right)^{-2} \int \psi^2(x-\vartheta_o)\, dF(x) ,$$

provided we can show that

$$(2.2.19) \quad n^{1/2} \sup_t \left| \frac{d}{ds}T(F^{(s)})\Big|_{s=t} - \frac{d}{ds}T(F^{(s)})\Big|_{s=0} \right| \longrightarrow 0$$

in probability.

We shall show (2.2.19) under the following additional assumption: There exists a neighbourhood U of $T(F) = \vartheta_o$ such that $u(x) = \sup \{ |\psi(x-\vartheta)| : \vartheta \in U \}$ defines a square integrable function in $L_2(F)$.

Let $b(t) = \int \psi'(x-T(F^{(t)}))\, dF^{(t)}(x)$. Then by the uniform convergence of $T(F^{(t)})$ to $T(F)$ (uniform in t)

$$(2.2.20) \qquad \sup_t | b(t) - b(0) | \longrightarrow 0 \quad \text{in probability.}$$

Let $U_k \searrow T(F)$ be a sequence of neighbourhoods of $T(F)$ contained in U. Since for every x

$$\sup_{\vartheta \in U_k} | \psi(x-\vartheta) - \psi(x-T(F)) | \searrow 0 \qquad \text{as } k \longrightarrow \infty$$

and since these functions are uniformly bounded by u it follows that

$$\lim_{k\to\infty} \int \left\{ \sup_{\vartheta\in U_k} \psi(x-\vartheta) - \psi(x-T(F)) \right\}^2 dF(x) = 0.$$

Let $\varepsilon > 0$. Choose k so large that the integral above is smaller than ε^3. Let n be so large that $P(T(F^{(t)}) \in U_k$ for all $0 \leq t \leq 1) > 1 - \varepsilon$. Then by Chebychev's inequality

$$P(n^{1/2} \sup_t | \int (\psi(x-T(F^{(t)})) - \psi(x-T(F)))\, d(F_n-F)(x) | \geq \varepsilon)$$

$$\leq \varepsilon + P(n^{1/2} | \int \sup_{\vartheta\in U_k} (\psi(x-\vartheta) - \psi(x-T(F)))\, d(F_n-F)(x) | \geq \varepsilon)$$

$$\leq 2\varepsilon.$$

Together with (2.2.20) this implies (2.2.19).

Example 2.2.7: (Example 2.1.9) continued)

Let $h : \mathbb{R}^m \longrightarrow \mathbb{R}$ be a symmetric kernel of degree m which is degenerate with respect to the continuous d.f. F. If T denotes the functional defined by

$$T(G) = \int \ldots \int h(x_1,\ldots,x_m) \prod_{i=1}^{m} dG(x_i)$$

as in Example 2.1.9, then the U-statistics $U_n(h)$ and T are related by (h vanishes on the diagonals w.l.o.g.)

$$T(F_n) = n^{-m} \binom{n}{m} m!\; U_n(h).$$

The derivatives of $T(F^{(t)})$ have been computed in Example 2.1.9. Because h is degenerate with respect to F the formula derived in Example 2.1.9 reduces to

$$\frac{d^p}{ds^p} T(F^{(s)})\Big|_{s=t} = m(m-1)\ldots(m-p+1)t^{m-p} \int \ldots \int h(x_1,\ldots,x_m) \prod_{j=1}^{m} d(F_n-F)(x_j)$$

so that $\frac{d^p}{ds^p} T(F^{(s)})\Big|_{s=0} = 0$ for $p = 1,\ldots,m-1$ and

$$= m! \int \ldots \int h(x_1,\ldots,x_m) \prod_{j=1}^{m} d(F_n-F)(x_j)$$

for $p = m$.

It follows from Corollary 2.2.1 that $n^{m/2}\,(T(F_n)-T(F))$ converges weakly to $Z(\psi)$ where $\psi(x_1,\ldots,x_m) = h(F^{-1}(x_1),\ldots,F^{-1}(x_m))$, provided $\|h\| < \infty$. Since h vanishes on the diagonals $\|h\|$ equals the $L_2(F^m)$-norm.

Consequently,

$$n^{m/2}\, U_n(h) \longrightarrow Z(\psi) \qquad \text{weakly as } n \longrightarrow \infty$$

provided that

$$\int \ldots \int h^2(x_1,\ldots x_m)\, dF(x_1)\ldots dF(x_m) < \infty .$$

This extends the weak convergence result on U-statistics in chapter 1 to degenerate kernels. Using the Decomposition Theorem for U-statistics it is easy to formulate a general convergence theorem for U-statistics (where h is not necessarily degenerate). We leave this as an exercise.

If $m = 2$, then by Proposition 2.2.1

$$Z(\psi) = \sum_{j \geq 1} \lambda_j \left\{ \left(\int f_j(x)\, dB(x)\right)^2 - 1 \right\}.$$

In particular, the χ^2- statistic in Example 2.1.4 can be modified to

$$\chi^2 = n^{-1} \sum_{i \neq j} (h(X_i, X_j) - 1)$$

where

$$h(x,y) = \begin{cases} p_j^{-1}(\alpha) & \text{if } a_j < x,y \leq a_{j+1} \\ 0 & \text{otherwise .} \end{cases}$$

In this case χ^2 converges weakly to

$$\sum_{j=1}^{r-1} \left(\int f_j(x)\, dB(x)\right)^2 - r + 1.$$

This follows repeating the arguments in Example 2.2.4 and we leave it as an exercise as well.

3. M-estimators

M-estimators were briefly introduced in Example 2.1.6 as a special case of differentiable statistical functionals. However, their general definition does not relate to this property of differentiability but merely to some monotonicity condition. On the other hand the method of proof for its asymptotic normality again needs some sort of differentiability.

The estimation of a location parameter in the location model $\{ F(\cdot,\vartheta) : \vartheta \in \Theta \}$, where $F(x,\vartheta) = F(x - \vartheta)$, usually is carried out by the least square estimator (minimize $\sum (x_i - \vartheta)^2$) or the maximum likelihood estimator (minimize $\sum \log f(x_i - \vartheta)$, Example 2.1.2). Instead of taking the specific convex functions $u \longrightarrow u^2$ or $u \longrightarrow \log f(u)$, Huber's M-estimator considers 'convex' functions more generally. The importance of this class of estimators may be seen in their robustness properties in quite a few examples. This means that the behaviour of the estimator still is reasonably good when the assumptions of the model are not exactly satisfied. For example, in the location model the d.f. F may vary in some small (open) set, or it may be contaminated by εG where $\varepsilon \leq \varepsilon_0$ and G is some known d.f. Classical estimators need not be robust. We shall not discuss the aspects of robustness for M-estimators, nor for R- and L-estimators considered in the next chapter. For this purpose the

reader may contact Huber's book on robust statistics.
In this section some of the basic mathematical properties of M-estimators are (briefly) discussed. We start with

Lemma 2.3.1: Let $\psi : \mathbb{R} \longrightarrow \mathbb{R}$ be non-decreasing and define for a d.f. F

$$(2.3.1) \qquad \lambda_F(t) = \int \psi(x - t)\, dF(x).$$

If for some t_o $|\lambda_F(t_o)| < \infty$, then λ_F is an everywhere defined, non-increasing function, possibly $\lambda_F(t) = \infty$ or $-\infty$. If ψ has positive (negative) values, so has λ_F.

Proof: If $s \geq t$, then $\psi(x-s) \leq \psi(x-t)$, in particular we obtain for $t \geq t_o$ that

$$\int \psi^+(x-t)\, dF(x) \leq \int \psi^+(x-t_o)\, dF(x) < \infty .$$

Analoguously for $t \leq t_o$

$$\int \psi^-(x-t)\, dF(x) \leq \int \psi^-(x-t_o)\, dF(x) < \infty .$$

It follows that λ_F is well defined, possibly infinity. Obviously λ_F also is non-increasing.

Definition 2.3.1: Let ψ be as in Lemma 2.3.1. A functional T defined on some subset of the d.f. is called an *M-estimator* (for a location parameter) (or a maximum likelihood type estimator) if

$$(2.3.2) \qquad \sup \{ t : \lambda_F(t) > 0 \} \leq T(F) \leq \inf \{ t : \lambda_F(t) < 0 \}$$

holds for every F where T is defined.
For empirical d.f. we may represent $T(F_n)$ as a measurable map $T_n : \mathbb{R}^n \longrightarrow \mathbb{R}$ satisfying

$$(2.3.3) \qquad T_*(x_1,\dots,x_n) = \sup \{ t : \lambda_{F_n}(t) > 0 \} \leq T_n(x) \leq$$
$$T^*(x_1,\dots,x_n) = \inf \{ t : \lambda_{F_n}(t) < 0 \}$$

where

$$F_n = F_n(x_1,\dots,x_n) = n^{-1} \sum_{i=1}^{n} \varepsilon_{x_i} ,$$

ε_{x_i} the point mass in x_i.

Lemma 2.3.2: (1) T_n is permutation invariant if it is a function of T^* and T_*.

(2) If T_n is a linear function of T^* and T_*, then it is equivariant with respect to the group $\mathbb{R}$ acting on $\mathbb{R}$ by $x \longrightarrow x + s$ $(s \in \mathbb{R})$ and on $\mathbb{R}^n$ by $(x_1,\ldots,x_n) \longrightarrow (x_1+s,\ldots, x_n+s)$ $(s \in \mathbb{R})$.

(3) If T_n is as in (2), then the family of distributions of $T_n(x_1+\vartheta,\ldots,x_n+\vartheta)$ under any probability P is a location model.

Proof: (1) Since F_n is invariant under permutations of the coordinates, T^* and T_* have the same property and so has T_n.

(2) Recall that a statistic $S : \mathbb{R}^n \longrightarrow \mathbb{R}$ is called equivariant with respect to the group G acting simultaneously on $\mathbb{R}$ and $\mathbb{R}^n$, if for every $x \in \mathbb{R}^n$ $S(gx) = gS(x)$, $g \in G$.

We shall show this for T_*. T^* is similar.

$$T_*(x_1+a,\ldots,x_n+a) = \sup\{ t : \lambda_{F_n(x_1+a,\ldots,x_n+a)}(t) > 0 \} =$$

$$= \sup\{ t : n^{-1} \sum_{i=1}^{n} \psi(x_i-t+a) > 0 \} =$$

$$= \sup\{ s+a : n^{-1} \sum_{i=1}^{n} \psi(x_i-s) > 0 \} =$$

$$= T_*(x_1,\ldots,x_n) + a.$$

Since T_n is a linear function of T^* and T_* (2) follows immediately.

(3) By (2) we have

$$P(x : T_n(x_1+\vartheta,\ldots,x_n+\vartheta) \leqq u) = P(x : T_n(x) \leqq u - \vartheta) =$$

$$= F(u-\vartheta),$$

where F denotes the d.f. of T_n under P.

Without proof let us recall two theorems (the Cramér-Rao inequality and Pitman's theorem) to illustrate the equivariance property.

Theorem 2.3.1: If ϑ is a location parameter, if $\hat{\vartheta}_n$ is an unbiased, equivariant estimator for ϑ and if the d.f. F is absolutely continuous with a differentiable density, then

$$\operatorname{Var} \hat{\vartheta}_n \geqq n^{-1} I(F)^{-1} .$$

Here

$$I(F) = \int \left(\frac{f'(x)}{f(x)} \right)^2 dF(x) \tag{2.3.4}$$

denotes the *Fisher information* of F where f is the density of F.

Theorem 2.3.2: Let ϑ be a location parameter and let U be an estimator of ϑ based on n i.i.d. observations $X_1,\dots,X_n$. Define Y by $Y = (X_2-X_1,\dots,X_n-X_1)$. Then

$$\hat{\vartheta}_n = U - E(U|Y)$$

minimizes the variance among all unbiased estimators which are equivariant with respect to the translation groups $x \longrightarrow x+s \quad (x,s \in \mathbb{R})$ and $x \longrightarrow x + (s,\dots,s) \quad (x \in \mathbb{R}^n,\ s \in \mathbb{R})$.

Example 2.3.1: Let $\psi(x) = 2x$. Then

$$\lambda_F(t) = n^{-1} \sum_{i=1}^{n} 2(x_i-t)$$

for $F = n^{-1} \sum_{i=1}^{n} \varepsilon_{x_i}$. Therefore $\lambda_F(t) = 0$ if $t = \frac{1}{n} \sum_{i=1}^{n} x_i$.

Since ψ is the derivative of x^2, this shows that the least square estimator is an M-estimator.

Example 2.3.2: Let K be fixed and ψ be the derivative of the map

$$t \longrightarrow \begin{cases} t^2/2 & \text{if } |t| \leq K \\ K|t| - K^2/2 & \text{if } |t| > K. \end{cases}$$

Then, setting

$$F = n^{-1} \sum_{i=1}^{n} \varepsilon_{x_i},$$

$$\lambda_F(t) = \int \psi(x-t)\, dF(x) = n^{-1} \sum_{i=1}^{n} \psi(x_i-t) =$$

$$= n^{-1} \sum_{i:|x_i-t|\leq K} (x_i-t) + \sum_{i:|x_i-t|>K} K \operatorname{sign}(x_i-t)$$

is very similar to the α-trimmed mean. The corresponding M-estimator 'disregards' small and large x-values.

Example 2.3.3: The maximum likelihood estimator for the location model $\{F(\cdot,\vartheta) : \vartheta \in \mathbb{R}\}$ with $dF(x) = f(x)\,dx$ minimizes $\sum \log f(X_i,\vartheta)$. Thus it is an M-estimator where ψ is given by

$$\psi(x) = -\frac{f'(x)}{f(x)}$$

Thus T(F) satisfies $\int f'(x - T(F))\, dx = 0$.

Example 2.3.4: We shall give an example of a different type of an M-estimator, that is an M-estimator for a scale parameter. Obviously this should be defined by

$$\int \psi\left(\frac{x}{T(F)} \right) dF(x) = 0.$$

The maximum likelihood estimator for a scale problem $\{F(\cdot,\sigma) : \sigma \in \Theta\}$ with d.f. $F(\cdot,\sigma)$ having densities $f(x,\sigma) = \sigma^{-1} f(x/\sigma)$ minimizes

$$\sum_{i=1}^{n} \log f(X_i,\sigma) .$$

Taking the partial derivative with respect to σ we find that

$$\psi(x) = - x \frac{f'(x)}{f(x)} - 1.$$

Another choice for ψ is

$$\psi(x) = \text{sign}(|x| - 1).$$

Then $T(F)$ estimates that point for which $F(z) - F(-z) = 1/2$ (the median absolute deviation from 0). Indeed

$$\begin{aligned} 0 = \int \psi(x/T(F))\, dF(x) &= 1 - F(T(F)) - (F(T(F)) - F(0)) + \\ &+ F(-T(F)) - (F(0) - F(-T(F))) = \\ = 1 + 2F(-T(F)) - 2F(T(F)) . \end{aligned}$$

The previous discussions, results and examples indicate the variety of applications of M-estimators only briefly. Here we are basically interested in the asymptotic behavior of M-estimators and we shall prove two results at the end of this section.

Theorem 2.3.3: Let λ_F be everywhere defined and let for some $t_o \in \mathbb{R}$

$$\lambda_F(t) > 0 \quad \text{if} \quad -\infty < t < t_o$$

and

$$\lambda_F(t) < 0 \quad \text{if} \quad t_o < t < \infty.$$

If $X_1, X_2, \ldots$ are independent, F-distributed random variables, then

$$\lim_{n\to\infty} T_n(X_1,\ldots,X_n) = t_o \quad \text{a.s.}$$

Proof: Let F_n denote the empirical distribution function of $X_1,\ldots,X_n$. Then

$$\lambda_{F_n}(t) = n^{-1} \sum_{i=1}^{n} \psi(X_i-t) \longrightarrow \int \psi(x-t)\, dF(x) = \lambda_F(t)$$

for all $t \in \mathbb{R}$ by the strong law of large numbers. If $t < t_o$ then $n^{-1} \sum \psi(X_i-t) > 0$ and consequently $T_n(X_1,\ldots,X_n) \geq t$.

Analoguously, for $t > t_o$ we have $T_n(X_1,\ldots,X_n) \leq t$. Both relations hold for fixed t and all n sufficiently large almost surely. Therefore, for every $\varepsilon > 0$

$$t_o - \varepsilon \leq \liminf_{n\to\infty} T_n(X_1,\ldots,X_n) \leq$$

$$\leq \limsup_{n\to\infty} T_n(X_1,\ldots,X_n) \leq t_o - \varepsilon \quad \text{a.s.}$$

Letting $\varepsilon \longrightarrow 0$ the theorem follows.

<u>Theorem 2.3.4:</u> Let λ_F be everywhere defined. Assume that for some $t_o \in \mathbb{R}$ $\lambda_F(t_o) = 0$ and that λ_F is strictly decreasing and continuous in a neighbourhood of t_o. Let

$$\int \psi^2(x - t)\, dF(x) < \infty$$

and let this function be continuous in $t = t_o$.
Denote by $T_n = T_n(X_1,\ldots,X_n)$ the M-estimators where $X_1, X_2, \ldots$ is an independent, F-distributed sequence of random variables. Then

$$n^{1/2}\, \lambda_F(T_n - t_o) \longrightarrow N(0,\sigma_1^2)$$

weakly where

$$\sigma_1^2 = \int \psi^2(x - t_o)\, dF(x) .$$

Moreover, if λ_F is continuously differentiable in a neighbourhood of t_o and if $\lambda_F'(t_o) \neq 0$, then

$$n^{1/2}\, (T_n(X_1,\ldots,X_n) - t_o) \longrightarrow N(0,\sigma^2)$$

weakly where

$$\sigma^2 = \sigma_1^2\, \lambda_F'(t_o)^{-2}.$$

<u>Proof:</u> W.l.o.g. we may assume that $t_o = 0$.
Fix $a \in \mathbb{R}$, $a \neq 0$. Choose $k_n \in \mathbb{R}$ satisfying

$$\lim_{n\to\infty} k_n = t_o = 0 \quad \text{and} \quad n^{1/2}\, \lambda_F(k_n) = a \quad (n \geq n(a)).$$

This is possible, since λ_F is monotone and continuous in some neighbourhood of 0 (define $k_n = \lambda_F^{-1}(an^{-1/2})$).
First observe the following properties:

(1) $\{ x : n^{1/2}\, \lambda_F(T_n(x)) \underset{(\geq)}{>} a\} = \{ x : T_n(x) \underset{(\leq)}{<} k_n \}$

since λ_F is strictly decreasing in a neighbourhood of all k_n.

(2) $\{ x : T_n(x) < k_n \} \subset \{x : \sum \psi(x_i - k_n) \leq 0 \}$

since by definition

$$n^{-1} \sum_{1\leq i\leq n} \psi(x_i - k_n) > 0 \quad \text{implies} \quad k_n \leq T_n(x).$$

(3) $\{x : \sum \psi(x_i - k_n) < 0\} \subset \{x : T_n(x) \leq k_n\}$

(this follows again by definition).

It follows from (1) - (3) that

$$(2.3.5) \quad P(n^{1/2}\lambda_F(T_n) > a) = P(T_n < k_n) \leq$$

$$\leq P(n^{-1/2}\sum_{i=1}^{n}(\psi(X_i-k_n) - \lambda_F(k_n)) \leq -a)$$

and

$$(2.3.6) \quad P(n^{1/2}\lambda_F(T_n) \geq a) = P(T_n \leq k_n) \geq$$

$$\geq P(n^{-1/2}\sum_{i=1}^{n}(\psi(X_i-k_n) - \lambda_F(k_n)) < -a).$$

Let us first assume that $\sigma_1^2 = 0$. Since $E(\psi(X_i-k_n)) = \lambda_F(k_n)$ and since $\tau^2(t) = \int(\psi(x-t) - \lambda_F(t))^2 \, dF(x)$ is continuous in $t = 0$,

$$E(n^{-1/2}\sum_{i=1}^{n}\psi(X_i-k_n) - \lambda_F(k_n))^2 \longrightarrow 0.$$

Therefore the right hand sides in (2.3.5) and (2.3.6) converge to 0 or 1 if $a > 0$ or if $a < 0$. Consequently,

$$n^{1/2}\lambda_F(T_n) \longrightarrow 0 \quad \text{in probability.}$$

Assume now that $\sigma_1^2 > 0$. In view of (2.3.5) and (2.3.6) it suffices to show that

$$n^{-1/2}\sum_{i=1}^{n} Y_{in} \longrightarrow N(0,\sigma_1^2)$$

where $Y_{in} = \psi(X_i-k_n) - \lambda_F(k_n) \quad (1 \leq i \leq n)$.

Observe that for each n $Y_{1n},\ldots,Y_{nn}$ are i.i.d. random variables. By assumption they have finite variances which are uniformly bounded from above and below. In order to show the Lindeberg condition it suffices to prove the following statement: For every $\varepsilon > 0$

$$\lim_{n\to\infty}\int_{\{|Y_{1n}| > \varepsilon n^{1/2}\}} Y_{1n}^2 \, dP = 0.$$

Since $\lambda_F(k_n) \longrightarrow 0$ this reduces again to

$$(2.3.7) \quad \lim_{n\to\infty}\int_{\{|\psi(X_1-k_n)| > \varepsilon n^{1/2}\}} \psi^2(X_1-k_n) \, dP = 0.$$

Assume that $a > 0$. Then by definition

$$\psi(X_1-k_N) \leq \psi(X_1-k_n) \leq \psi(X_1)$$

for all $n \geq N$ provided N is large enough. It follows that

$$\psi^2(X_1-k_n) \leq \max\{\psi^2(X_1), \psi^2(X_1-k_N)\}$$

and this inequality holds also if $a < 0$. Keeping N fixed we see that $\{\psi^2(X_1-k_n) : n \geq N\}$ is uniformly integrable and hence (2.3.7) follows.

The Lindeberg condition implies that $n^{-1/2} \sum_{i=1}^{n} Y_{in}$ converges weakly to some normal variable with expectation 0 and variance $\lim_{n\to\infty} \text{Var } Y_{1n} = \lim_{n\to\infty} \tau^2(k_n) = \sigma_1^2$. This proves that

$$\lim_{n\to\infty} P(n^{1/2} \lambda_F(T_n) \geq \sigma_1 a) = \frac{1}{(2\pi)^{1/2}} \int_a^\infty e^{-t^2/2} dt$$

and since a was chosen arbitrarily the first part of the theorem follows.

For the second part of the theorem write

$$\lambda_F(T_n) = \lambda_F(0) + \lambda_F'(\xi)\, T_n$$

where $|\xi| \leq |T_n|$. Since by Theorem 2.3.3 $T_n \longrightarrow 0$ a.s. we also have $\lambda_F'(\xi) \longrightarrow \lambda_F'(0)$ a.s. by continuity. Therefore

$$\frac{n^{1/2}\, \lambda_F(T_n)}{n^{1/2}\, \lambda_F'(0)\, T_n} \longrightarrow 1 \quad \text{a.s.}$$

and $n^{1/2}\, T_n$ converges weakly to $N(0,\sigma^2)$.

Notes on chapter 2:

The definition of a von Mises' functional goes back to von Mises in 1947. In this paper he also obtained Corollary 2.2.1 for m=2. The approach presented in section 2 is due to Filippova (1962) where one also can find sufficient conditions for a functional to be a von Mises' functional. Very much in the same way as the distribution invariance principles were proven in the first chapter, Filippova's approach can be extended to distribution invariance principles for degenerate U- and V-statistics replacing the Brownian bridge by Müller-Kiefer processes. This has been done by Grillenberger, Keller and the author (1981). Almost sure invariance principles in the degenerate case (including the law of the iterated logarithm) recently have been investigated by Dehling, Philipp and the author (1983). Also recently, Fernholz (1983) published her work on von Mises' calculus from which Propositions 2.1.1 and 2.1.2 are taken. In this book one can find applications of the propositions, for example to M-, R- and L-estimators. In chapter 3 differentiability plays an important role again. Finally, for the results on M-estimators we refer to Huber's book (1981).

CHAPTER 3: STATISTICS BASED ON RANKING METHODS

Central to the material presented in this chapter is Theorem 3.4.2 below. Although it is formulated for linear rank statistics, the idea of proof will appear in sections 3.5 and 3.6 and chapter 4 again. It provides a general viewpoint for rank statistics, signed rank statistics and linear combinations of a function of the order statistic (and some more which are not discussed here). Rank statistics are considered in some detail in sections 2-4, and since they are a special case of permutation tests, we prove the famous Wald-Wolfowitz-Noether-Hájek theorem in section 1.

Statistics based on ranking methods play the most important role in applications of nonparametric statistics to practical problems. They are easy to compute, simple in structure and have good optimality and efficiency properties compared with parametric models. Their main advantage may be seen from the fact that they do not rely so much on the exact knowledge of the underlying distribution. Although they were originally proposed as analoga of finite sample procedures for parametric models (cf. the Wilcoxon tests in chapter 1), their "asymptotic distribution-free" property justifies (for the experimenter) the application for practical problems even if their exact distribution is not known or not computable. In many cases, even for small sample sizes, the approximation is sufficiently good.

1. Permutation tests

Permutation tests have first been discussed by Fisher in 1935 and later in a more complete fashion by Pitman. The first limit theorem is due to Wald and Wolfowitz in 1944, extended subsequently by Hoeffding, Noether, Dwass, Frazer and Motoo. Below we shall prove the asymptotic normality as in Hájek (1961).

Recall that γ_n denotes the permutation group of n elements acting on $\mathbb{R}^n$ by $x = (x_1,\dots,x_n) \longrightarrow \tau(x) = (x_{\tau(1)},\dots,x_{\tau(n)})$ for $\tau \in \gamma_n$.

Definition 1.3.1: A measurable map $f : \mathbb{R}^n \longrightarrow \mathbb{R}$ is called a *permutation statistic* if there exists an $\alpha \in (0,1)$ such that

$$\frac{1}{n!} \sum_{\tau\in\gamma_n} f(\tau(x)) = \alpha$$

for all $x \in \mathbb{R}^n$. A test with critical function f is called a *permutation test*.

Consider the family H_o of all n-fold product measures of absolutely continuous distribution functions. Since this family is complete and sufficient, the theorem of Lehmann and Stein (cf. Lehmann 1959, p. 184) asserts that the family of permutation tests coincides (under H_o) with all similar tests (i.e. $\int \varphi \, dF = \alpha$ for some $\alpha \in (0,1)$ and for all F belonging to H_o).

We begin with two examples.

Example 3.1.1: Let $a(i,j)$ $(1 \leq i,j \leq n)$ be a matrix of real numbers. A statistic

$$T(x) = \sum_{i=1}^{n} a(i,r_i(x)) \qquad (x \in \mathbb{R}^n)$$

is called a *linear rank statistic* where $r_i(x)$ denotes the rank of x_i among $x_1,\dots,x_n$.

For a permutation $\tau \in \gamma_n$ one has $r_i(\tau(x)) = r_j(x)$ iff $\tau(i) = j$ and therefore

$$\frac{1}{n!} \sum_{\tau\in\gamma_n} T(\tau(x)) = \frac{1}{n!} \sum_{\tau\in\gamma_n} \sum_{i=1}^{n} a(i,r_i(\tau(x))) =$$

$$= \frac{1}{n} \sum_{i=1}^{n} \sum_{j=1}^{n} a(i,r_j(x)) = \frac{1}{n} \sum_{i=1}^{n} \sum_{j=1}^{n} a(i,j)$$

is independent of $x \in \mathbb{R}^n$. This type of statistic will be considered in detail in the following sections.

U- and V-statistics are in general no permutation statistics; however, in special cases like the Wilcoxon two-sample statistic they are of this type.

Example 3.1.2: Let $b_1,\ldots,b_n,a_1,\ldots,a_n$ be real numbers such that $a_i \neq a_j$ for $i \neq j$, and let $X_1,\ldots,X_n$ be random variables satisfying $P(X_i = a_{\tau(i)}, 1 \leq i \leq n) = n!^{-1}$ for every $\tau \in \gamma_n$. Then the statistic $T = \sum_{1\leq j\leq n} b_j X_j$ is called a *linear permutation* ***statistic.***

It is easy to see that T is a.s. a permutation statistic.

Lemma 3.1.1:

Let $T = \sum_{j=1}^{n} b_j X_j$ be a linear permutation statistic where the random vector $X = (X_1,\ldots,X_n)$ assumes a.s. only the values of the permutations of the fixed vector $a = (a_1,\ldots,a_n)$ with equal probabilities. Then

$$(3.1.1) \qquad ET = n\,\bar{a}\,\bar{b}$$

and

$$(3.1.2) \qquad \operatorname{Var} T = \frac{1}{n-1} \sum_{j=1}^{n} (b_j-\bar{b})^2 \sum_{i=1}^{n} (a_i-\bar{a})^2$$

where $\bar{a} = n^{-1} \sum_{i=1}^{n} a_i$ and $\bar{b} = n^{-1} \sum_{i=1}^{n} b_i$.

Proof:

(3.1.1) is easy: since for all permutations $\tau \in \gamma_n$ $P(X=\tau a) = n!^{-1}$,

$$ET = n!^{-1} \sum_{\tau} \sum_{j=1}^{n} b_j a_{\tau(j)} = \sum_{j=1}^{n} b_j\, n!^{-1} \sum_{\tau} a_{\tau(j)} = n\,\bar{b}\,\bar{a}\,.$$

(3.1.2) follows from

$$\operatorname{Var} T = E\Big(\sum b_i X_i - n\bar{b}\bar{a}\Big)^2 =$$

$$= n!^{-1} \sum_{\tau} \Big(\sum_{i=1}^{n} b_i (a_{\tau(i)}-\bar{a})\Big)^2$$

$$= n^{-1} \sum b_i^2 \sum (a_j-\bar{a})^2 - (n(n-1))^{-1} \sum_{k\neq i} b_k b_i \sum (a_j-\bar{a})^2$$

$$= (n-1)^{-1} \sum (a_j-\bar{a})^2 \sum (b_i-\bar{b})^2.$$

For the Wald-Wolfowitz-Noether-Hájek theorem on the asymptotic normality we shall use a normalized formulation. Let $T_N = \sum_{j=1}^{N} b_j X_{jN}$ $(N \geq 1)$ be a sequence of permutation statistics where for each N $X_N = (X_{1N},\ldots,X_{NN})$ is a random vector with permutation-invariant distribution and $P(X_N=a_N) = N!^{-1}$ for some sequence $a_N = (a_{1N},\ldots,a_{NN})$ with $a_{iN} \neq a_{jN}$ $(1\leq i\neq j\leq N)$. Thus, $E\,T_N$ and $\mathrm{Var}\,T_N$ are given by (3.1.1) and (3.1.2) respectively. As far as the limit distribution of T_N is concerned, we may assume that $\bar{a}_N = \bar{b}_N = 0$ and $\sum a_{iN}^2 = \sum b_{iN}^2 = N$. In fact, setting

$$\tilde{a}_N = N^{-1/2}(\sum(a_{iN}-\bar{a}_N)^2)^{1/2}$$

and

$$\tilde{b}_N = N^{-1/2}(\sum(b_{iN}-\bar{b}_N)^2)^{1/2}$$

we obtain

$$\sum \tilde{b}_N^{-1}(b_{jN}-\bar{b}_N)\tilde{a}_N^{-1}(X_{jN}-\bar{a}_N) = (\tilde{a}_N\tilde{b}_N)^{-1}(T_N-N\bar{a}_N\bar{b}_N)$$

provided $\tilde{a}_N \neq 0$ and $\tilde{b}_N \neq 0$. (If $\tilde{a}_N$ or $\tilde{b}_N = 0$ T_N reduces to a constant.)

With these assumptions and notations we have

Theorem 3.1.1:
The following two statements are equivalent:

(3.1.3)

(a) $N^{-1/2}T_N$ converges weakly to $\mathcal{N}(0,1)$.

(b) $\lim_{N\to\infty} \max_{1\leq i\leq N} N^{-1}a_{iN}^2 = 0.$

(c) $\lim_{N\to\infty} \max_{1\leq i\leq N} N^{-1}\, b_{iN}^2 = 0.$

(3.1.4) $$\lim_{N\to\infty} \sum_{\substack{i,j=1 \\ |a_{iN}b_{jN}|\geq \varepsilon N^{1/2}}}^{N} N^{-2}\, a_{iN}^2 b_{jN}^2 = 0$$

for any $\varepsilon > 0$.

Proof:
The proof is divided into two parts:

(A) If the X_{iN} are i.i.d. (the X_{iN} in (3.1.3) (a) satisfy $P(X_{iN} = a_{jN}) = N^{-1}$) then the equivalence of (3.1.3) and (3.1.4) is just the central limit theorem under the Lindeberg condition.

(B) Given N, there exists a version of X_N and i.i.d. random variables $Y_{1N},\dots,Y_{NN}$ with $P(Y_{jN} = a_{iN}) = N^{-1}$ defined on the same probability space, such that

$$N^{-1/2} \sum b_{iN}(X_{iN}-Y_{iN}) \longrightarrow 0$$

in L_2, provided (3.1.3)(b) or (3.1.4) holds.

Proof of (A): Let $Y_{1N},\dots,Y_{NN}$ be i.i.d. random variables with $P(Y_{1N} = a_{jN}) = N^{-1}$ $(N \geq 1)$. Then $\{N^{-1/2} b_{iN} Y_{iN} : 1 \leq i \leq N, N \geq 1\}$ is an array of random variables with zero expectation and variance $N^{-1} b_{iN}^2 N^{-1} \sum a_{jN}^2 = N^{-1} b_{iN}^2$.

Assume that (3.1.3) holds, in particular $\sum_{i=1}^{N} N^{-1/2} b_{iN} Y_{iN}$ converges weakly to $\mathcal{N}(0,1)$. From (c) it follows that

$$P(|N^{-1/2} b_{iN} Y_{iN}| > \varepsilon) \leq \varepsilon^{-2} N^{-1} b_{iN}^2 \longrightarrow 0 \quad \text{uniformly in}$$

$1 \leq i \leq N$ as $N \longrightarrow \infty$. Therefore, the Lindeberg condition is satisfied for the array under consideration, that is (3.1.4):

$$\sum_{j=1}^{N} \int_{\{|N^{-1/2} b_{jN} Y_{jN}| \geq \varepsilon\}} N^{-1} b_{jN}^2 Y_{jN}^2 \, dP = \sum_{\substack{i,j=1 \\ |N^{-1/2} a_{iN} b_{jN}| \geq \varepsilon}}^{N} N^{-2} b_{jN}^2 a_{iN}^2 \to 0.$$

Conversely assume that (3.1.4) holds. Then the Lindeberg condition is satisfied and thus (3.1.3)(a) follows. Because of

$$N^{-1} b_{iN}^2 = E(N^{-1/2} b_{iN} Y_{iN})^2 \leq \varepsilon^2 + \int_{\{|N^{-1/2} b_{iN} Y_{iN}| \geq \varepsilon\}} N^{-1} b_{iN}^2 Y_{iN}^2 \, dP \to 0$$

uniformly in $i \in \{1,\dots,N\}$ as $N \longrightarrow \infty$ and $\varepsilon \longrightarrow 0$, (3.1.3)(c) holds. Finally, (b) is a consequence of

$$(3.1.5) \qquad N^{-1} a_{iN}^2 = N^{-2} \sum_{j=1}^{N} a_{iN}^2 b_{jN}^2 \leq \varepsilon^2 + \sum_{\substack{j=1 \\ |a_{iN} b_{jN}| \geq \varepsilon N^{1/2}}}^{N} N^{-2} a_{iN}^2 b_{jN}^2 \xrightarrow[\varepsilon \to 0]{N \to \infty} 0.$$

Proof of (B): We fix $N \in \mathbb{N}$ and we shall omit the index N. Let $\Omega = \mathbb{R}^N \times \mathbb{R}^N$. We have to define a probability P on Ω such that its first marginal is the distribution of X_N and its second one that of Y_N, defined in (A). Write $a = a_N = (a_1,\dots,a_N)$ with $a_1 < a_2 < \dots < a_N$.

If $c = \{c_1 < \dots < c_m\} \subset \{a_1,\dots,a_N\}$ and $t = (t_1,\dots,t_m) \in \mathbb{N}^m$ with $\sum t_j = N$ are given, define

$$d = d(c,t) = (\underbrace{c_1,\dots,c_1}_{t_1\text{-times}},\dots,\underbrace{c_m,\dots,c_m}_{t_m\text{-times}})$$

and

$$(3.1.6) \qquad P(\{\tau a,\tau' d\}) = \begin{cases} N^{-N} \prod_{j=1}^{m} t_j!^{-1} & \text{if } \tau d=\tau' d,\ \tau,\tau' \in \gamma_N \\ 0 & \text{otherwise.} \end{cases}$$

This defines a probability on Ω. Let us first observe that P has the correct marginals. If $u \in \gamma_N d$ is fixed

$$P(\mathbb{R}^N \times \{u\}) = \sum_{\tau \in \gamma_N} P(\{(\tau a,u)\}) =$$

$$= \sum_{\tau \in \gamma_N,\, \tau d = u} P(\{(\tau a,u)\}) = N^{-N}$$

and if $\tau \in \gamma_N$ is fixed, then

$$P(\{\tau a\} \times \mathbb{R}^N) = \sum_{u \in \{a_1,\dots,a_N\}^N} P(\{(\tau(a),u)\}) =$$

$$= \sum_d P(\{(\tau(a),\tau(d))\}) = \sum_d N^{-N} \prod t_j!^{-1}$$

$$= \sum_{\substack{t_1,\dots,t_m \\ \Sigma t_j = N}} N^{-N} \prod_{j=1}^{m} t_j!^{-1} \binom{N}{m} = N!^{-1}.$$

(The last equality holds because $\sum N! \prod t_j!^{-1} \binom{N}{m}$ is the number of elements in $\{1,\dots,N\}^N$.)

It is left to show that

$$N^{-1/2} \sum b_i (X_i - Y_i) \longrightarrow 0 \quad \text{in } L_2$$

and in fact we shall show that

$$
\begin{aligned}
&N^{-1}\, E\Big(\sum b_i (X_i - Y_i)\Big)^2 \le 2E(X_1 - Y_1)^2 \\
(3.1.7)\qquad & \\
&\le 8\big(N^{-1} \max_{1\le i\le N} a_i^2 + N^{-1/2} \max_{1\le i\le N} |a_i|\big).
\end{aligned}
$$

By (3.1.5), if (3.1.3) or (3.1.4) holds, the right-hand side always tends to zero.

Since the probability P defined in (3.1.6) is invariant under the action of $\tau \times \tau$ $(\tau \in \gamma_N)$, the first inequality in (3.1.7) follows from

$$
\begin{aligned}
N^{-1}\, E\Big(\sum b_i (X_i - Y_i)\Big)^2 &= N^{-1} \sum_{i,j} b_i b_j\, E(X_i - Y_i)(X_j - Y_j) \\
&= N^{-1} \sum_i b_i^2\, E(X_1 - Y_1)^2 + N^{-1} \sum_{i \neq j} b_i b_j\, E(X_1 - Y_1)(X_2 - Y_2) \\
&= E(X_1 - Y_1)^2 - N^{-1} \sum_i b_i^2\, E(X_1 - Y_1)(X_2 - Y_2) \le 2\, E(X_1 - Y_1)^2 .
\end{aligned}
$$

It is left to show the second inequality in (3.1.7). Define random variables V_i by $V_i(\tau(a), \tau(d)) = d_i$ where $d = d(c,t) = (d_1, \ldots, d_N)$ as before in (3.1.6). We have $V_i = Y_1$ iff $X_1 = a_i$ and therefore

$$
\begin{aligned}
E(X_1 - Y_1)^2 &= N^{-1} \sum_{i=1}^{N} E(X_i - Y_i)^2 = N^{-1} \sum_{i=1}^{N} \sum_{j=1}^{N} \int_{\{V_j = Y_i\}} (X_i - Y_i)^2\, dP \\
(3.1.8)\qquad &= N^{-1} \sum_{j=1}^{N} \sum_{i=1}^{N} \int_{\{X_i = a_j\}} (a_j - V_j)^2\, dP \\
&\le 2\, N^{-1} \Big[\sum_{j=1}^{N} E(V_j^+ - a_j^+)^2 + \sum_{j=1}^{N} E(V_j^- - a_j^-)^2 \Big].
\end{aligned}
$$

We shall estimate each summand separately, but because of symmetry it suffices to consider $E(V_j^+ - a_j^+)^2$ only.

Let $\varepsilon_k = 1_{(a_k^+, \infty)}$ and $W_k := \operatorname{card}\{j : Y_j \le a_k^+\}$ $(k \ge 1)$.

We first claim that

$$
(3.1.9)\qquad \sum_{i=1}^{N} [\varepsilon_k(V_i^+) - \varepsilon_k(a_i^+)]^2 \le |W_k - k| + 1 \qquad (1 \le k \le N).
$$

In order to prove (3.1.9) fix $1 \leq k \leq N$, $\tau \in \gamma_N$ and $d = d(c,t) = (d_1,\dots,d_N)$. Denote by L that index satisfying $d_L \leq a_k^+ < d_{L+1}$. Then $V_i^+ \leq a_k^+$ iff $d_i \leq a_k^+$, i.e. $i \leq L$ and hence on $(\tau(a),\tau(d))$

$$(\varepsilon_k(V_i^+) - \varepsilon_k(a_i^+))^2 \begin{cases} = 0 & \text{if } i \leq \min(k,L) \text{ or if } \\ & i \geq \max(k+1,L+1) \\ \leq 1 & \text{otherwise.} \end{cases}$$

Since $\min(k,L)+N - \max(k+1,L+1) = N - |L-k| - 1$, we obtain

$$\sum_{i=1}^{N} (\varepsilon_k(V_i^+) - \varepsilon_k(a_i^+))^2 \leq |L-k| + 1 \quad \text{on} \quad (\tau(a),\tau(d)).$$

(3.1.9) follows now because $W_k((\tau(a),\tau(d))) = L$.

Let us now return to the estimation of the right-hand side in (3.1.8). Using the equality

$$a_i^+ = \sum_{k=1}^{N-1} (a_{k+1}^+ - a_k^+)\varepsilon_k(a_i^+),$$

the inequality $(l \leq k)$

$$(\varepsilon_k(V_i^+)-\varepsilon_k(a_i^+))(\varepsilon_l(V_i^+)-\varepsilon_l(a_i^+)) \leq (\varepsilon_k(V_i^+)-\varepsilon_k(a_i^+))^2$$

and (3.1.9) we conclude that

$$N^{-1} \sum_{j=1}^{N} E(V_j^+ - a_j^+)^2 = N^{-1} \sum_j E(\sum_{k=1}^{N-1} (a_{k+1}^+ - a_k^+)(\varepsilon_k(V_j^+)-\varepsilon_k(a_j^+)))^2$$

$$\leq N^{-1} \sum_{k,l=1}^{N-1} (a_{k+1}^+ - a_k^+)(a_{l+1}^+ - a_l^+)[E|W_{\max(k,l)} - \max(k,l)| + 1].$$

Since $W_i = \sum_{u=1}^{N} 1_{(-\infty,a_i^+]}(Y_u)$ has a Bernoulli distribution with $P(Y_1 \leq a_i^+) = iN^{-1}$ we have

$$E|W_i - i| \leq (i(1-iN^{-1}))^{1/2}.$$

We derive (3.1.7) using Cauchy-Schwarz' inequality and

$$N \geq \sum a_i^{+2} = \sum_{k,l=1}^{N-1} (a_{k+1}^+ - a_k^+)(a_{l+1}^+ - a_l^+)(N-\max(k,l)) \; :$$

$$N^{-1} \sum_{j=1}^{N} E(V_j^+ - a_j^+)^2 \le N^{-1} \left(\sum_{k=1}^{N-1} (a_{k+1}^+ - a_k^+)\right)^2$$

$$+ N^{-1} \left| \sum_{k=1}^{N-1} (a_{k+1}^+ - a_k^+) \right| \left(\sum_{k,l=1}^{N-1} (a_{k+1}^+ - a_k^+)(a_{l+1}^+ - a_l^+)(N - \max(k,l)) \right)^{1/2}$$

$$\le N^{-1} a_N^2 + N^{-1/2} |a_N| .$$

Remark:

According to Lemma 3.1.1, the variance of T_N in Theorem 3.1.1 is given by $N^2/(N-1)$. Hence, as $N \to \infty$, $\operatorname{Var} N^{-1/2} T_N \to 1$.

2. Simple linear rank statistics

Linear rank statistics were defined in Example 3.1.1. As a special case of a permutation statistic we can define rank statistics as follows.

Definition 3.2.1:

A statistic $T : \mathbb{R}^N \to R$ which can be written as a function of the ranks is called a *rank statistic*. T is a *linear rank statistic* if it is of the form

$$(3.2.1) \qquad T(x) = \sum_{i=1}^{N} a(i, r_i)$$

where $a(i,j)$ is a given matrix and where r_i denotes the rank of x_i among all coordinates $x_1, \dots, x_N$ of $x \in \mathbb{R}^N$. If the matrix $\{a(i,j)\}$ can be expressed as

$$a(i,j) = C_i \, a(j) \qquad (1 \le i, j \le N)$$

where C_i and $a(j)$ are constants, then T is said to be a *simple linear rank statistic*. The constants C_i are called *regression constants* and the $a(j)$ *scores*.

Similarly to Lemma 3.1.1 we first observe

Lemma 3.2.1:

Let $T = \sum_{i=1}^{N} a(i, r_i)$ be a linear rank statistic and let P be a permutation-invariant probability measure on $\mathbb{R}^N$ without atoms. Then, under P,

(3.2.2) $ET = N\bar{a}$

and

(3.2.3) $\text{Var } T = (N-1)^{-1} \sum_{i=1}^{N} \sum_{j=1}^{N} [a(i,j)-\bar{a}(\cdot,j)-\bar{a}(i,\cdot)+\bar{a}]^2$

where

$$\bar{a}(\cdot,j) = \frac{1}{N} \sum_{i=1}^{N} a(i,j), \bar{a}(i,\cdot) = \frac{1}{N} \sum_{j=1}^{N} a(i,j), \ \bar{a} = \frac{1}{N^2} \sum_{i,j=1}^{N} a(i,j).$$

Proof:

Since $P(r_i=j) = \frac{1}{N}$ $(1 \leq j \leq N)$, $E\, a(i,r_i) = \bar{a}(i,\cdot)$ and (3.2.2) follows. In order to show (3.2.3) note that

$$P(r_i=k, r_j=l) = \begin{cases} (N(N-1))^{-1} & \text{if } i \neq j \text{ and } k \neq l \\ 0 & \text{if } i \neq j \text{ and } k = l \\ N^{-1} & \text{if } i = j \text{ and } k = l. \end{cases}$$

Therefore

$$E\, T^2 = N^{-1} \sum_{i,j=1}^{N} a(i,j)^2 + (N(N-1))^{-1} \sum_{i \neq j} \sum_{k \neq l} a(i,k)a(j,l)$$

and after some routine calculations (3.2.3) follows. (Exercise.)

We shall now concentrate on simple linear rank statistics and first prove an optimality result.

Definition 3.2.2: Let $\{P_\vartheta : \vartheta \in I\}$ be a family of distributions where $I \subset \mathbb{R}$ is some interval containing 0. A test φ is called *locally most powerful* for $H = \{P_0\}$ against $K = \{P_\vartheta : \vartheta > 0\}$ at level α if it is uniformly most powerful at level α for H against $\{P_\vartheta : 0 < \vartheta < \varepsilon\}$ for some $\varepsilon > 0$. This means formally, that $\int \varphi \, d\, P_0 \leq \alpha$ and

$$\int \varphi \, dP_\vartheta \geq \int \varphi' \, d\, P_\vartheta \quad (0 < \vartheta < \varepsilon)$$

for every other level α test φ'.

Theorem 3.2.1:

Let $\{P_\vartheta : \vartheta \in I\}$ be a family of distributions on $\mathbb{R}$ where I is some open interval containing 0. Assume that each P_ϑ is absolutely continuous with density $p(\vartheta,\cdot)$ satisfying

(3.2.4) For almost all x, $\vartheta \to p(\vartheta,x)$ is differentiable in some neighbourhood of 0 (independent of x) with derivative $p'(\vartheta,x) = \frac{\partial}{\partial\vartheta}[p(\vartheta,x)]$

and

(3.2.5) $$\lim_{\vartheta\to 0} \int_{-\infty}^{\infty} |p'(\vartheta,x)|dx = \int_{-\infty}^{\infty} |p'(0,x)|dx < \infty.$$

For (fixed) $N \in \mathbb{N}$ let $T = \sum_{i=1}^{N} c_i\, a(r_i)$ be the simple linear rank statistic defined by regression constants $c_i \in \mathbb{R}$ and scores

(3.2.6) $$a(i) = E\left\{\frac{p'(0,X_{(i)})}{p(0,X_{(i)})}\right\}$$

where $\{X_{(i)} : 1 \leq i \leq N\}$ denotes the order statistic from an independent sample of size N drawn from the distribution P_0. Then, given $\alpha > 0$, there exists a $k = k(\alpha)$ such that the test with critical region $\{T>k\}$ is a locally most powerful rank test for $H = \{Q_0 = P_0^N\}$ against $K = \{Q_\vartheta = \prod_{i=1}^{N} P_{\vartheta c_i} : \vartheta>0\}$ at level α.

<u>Proof:</u> Denote by $r(x) = (r_1(x),\ldots,r_N(x))$ the rank vector of x. Fix an alternative $\vartheta > 0$. Since H is simple it follows from the Neyman-Pearson Lemma that the most powerful rank test φ_0 for H against ϑ has the form $\{\varphi_0 = 1\} = \{r : Q_\vartheta(r) > K/N!\}$ where K is determined by α : $N!\alpha - 1 < |\{r : Q_\vartheta(r) > K/N!\}| \leq \alpha N!$. Hence it has to be shown that for different ϑ (but small enough) the same rank vectors form the critical region and that these vectors can be described by the requirement that $\sum c_i\, a(r_i) > k(\alpha)$. This will follow at once, if we can show the following: There exists an $\varepsilon > 0$ such that for any two rank vectors r and r' the inequality

$$\sum c_i\, a(r_i) > \sum c_i\, a(r_i')$$

implies $Q_\vartheta(\{x : r(x) = r\}) > Q_\vartheta(\{x : r(x) = r'\})$ for all $\vartheta < \varepsilon$.

Let $A = \{x : r(x) = r\}$. Then for any ϑ

$$Q_\vartheta(A) = Q_o(A) + (Q_\vartheta - Q_o)(A)$$

$$= N!^{-1} + \vartheta \sum_{k=1}^{N} \int \dots \int_A \vartheta^{-1}(p(\vartheta C_k, x_k) - p(0,x_k)) \prod_{i=1}^{k-1} p(\vartheta C_i, x_i) \prod_{j=k+1}^{N} p(0,x_j)\, dx_1 \dots dx_N.$$

For fixed k define

$$g_\vartheta(x_1,\dots,x_N) = \vartheta^{-1}(p(\vartheta C_k, x_k) - p(0,x_k)) \prod_{i=1}^{k-1} p(\vartheta C_i, x_i) \prod_{j=k+1}^{N} p(0,x_j)$$

and

$$g(x_1,\dots,x_N) = C_k\, p'(0,x_k) \prod_{i \neq k} p(0,x_i).$$

The claim will follow if for each k

$$(3.2.7) \qquad \lim_{\vartheta \to 0} \int_A g_\vartheta \, d\lambda^N = \int_A g \, d\lambda^N \qquad (\lambda^N \text{ Lebesgue measure}).$$

Indeed, since Q_o is permutation-invariant and since for $\tau \in \gamma_N$ $r_k((x_{\tau(1)},\dots,x_{\tau(n)})) = r_{\tau(k)}(x)$ $(x_{\tau(k)} = x_{(r_k(\tau(x)))})$

$$C_k^{-1} \int_A g \, d\lambda^N = \int_A \frac{p'(0,x_k)}{p(0,x_k)} \, dQ_o(x_1,\dots,x_N)$$

$$= \frac{1}{N!} \sum_{\tau \in \gamma_N} \int_{\tau^{-1}A} \frac{p'(0,x_{\tau(k)})}{p(0,x_{\tau(k)})} \, dQ_o(x)$$

$$= \frac{1}{N!} \sum_{\tau \in \gamma_N} \int_{\tau^{-1}A} \frac{p'(0,x_{(r_k)})}{p(0,x_{(r_k)})} \, dQ_o(x) =$$

$$= \frac{1}{N!} E\left\{ \frac{p'(0,X_{(r_k)})}{p(0,X_{(r_k)})} \right\}$$

$$= \frac{1}{N!} a(r_k).$$

Thus it is left to show (3.2.7) for a fixed k . By Fubini's theorem

$$\int |g_\vartheta| \, d\lambda^N = \vartheta^{-1} \int_{-\infty}^{\infty} \left| \int_0^{\vartheta c_k} p'(\delta,x)\,d\delta \right| dx$$
$$\leq \vartheta^{-1} \int_0^{\vartheta c_k} \int_{-\infty}^{\infty} |p'(\delta,x)| \, dx \, d\delta$$

and hence by (3.2.5)

$$(3.2.8) \qquad \limsup_{\vartheta \to 0} \int |g_\vartheta| \, d\lambda^N \leq c_k \int_{-\infty}^{\infty} |p'(0,x)| \, dx = \int |g| \, d\lambda^N.$$

Since $g_\vartheta \longrightarrow g$ λ^N-almost surely as $\vartheta \to 0$, Fatou's lemma gives

$$\liminf \int_B g_\vartheta^\pm \, d\lambda^N \geq \int_B g^\pm \, d\lambda^N$$

for each measurable set B . If for some sequence $\vartheta_k \to 0$ and some B, $\lim_{k\to\infty} \int_B g_{\vartheta_k}^+ \, d\lambda^N > \int_B g^+ \, d\lambda^N$, it follows that $\limsup \int |g_{\vartheta_k}| \, d\lambda^N > \int |g| \, d\lambda^N$, contradicting (3.2.8). Hence $\lim_{\vartheta\to 0} \int_B g_\vartheta^+ \, d\lambda^N = \int_B g^+ \, d\lambda^N$. Interchanging $+$ and $-$, (3.2.7) follows.

We shall close this section with examples of simple linear rank statistics, which are of some importance.

<u>Example 3.2.1:</u> The Wilcoxon two-sample statistic

$$W = \sum_{i=1}^{n} R_i$$

(cf. Example 1.4.1) may be considered as a simple linear rank statistic with regression constants $c_i = \begin{cases} 1 & i \leq n \\ 0 & i > n \end{cases}$ and scores $a(i) = i$. Since this statistic is based on two independent samples $X_1,\ldots,X_n$ and $Y_1,\ldots,Y_m$ of i.i.d. random variables, it is a special case of a *two-sample linear rank statistic* $T(x_1,\ldots,x_n,y_1,\ldots,y_m) = \sum_{i=1}^{n} a(r_i)$, where r_i denotes the rank of x_i among $x_1,\ldots,x_n,y_1,\ldots,y_m$. Such a statistic may also be represented in the following form:

Define the random variables $V_i = 0$ or 1 if the i-th largest random variable in the combined sample originally was a Y- or X-variable $(1 \le i \le N = n+m)$. Then

$$T = \sum_{i=1}^{n+m} a(i)\, V_i \ .$$

If the distribution F of X_1 equals the distribution G of Y_1, then the distribution of $(V_1,\ldots,V_N)$ obviously is permutation-invariant. Thus, in order to compute the moments and correlation, one obtains $E\,V_i = \frac{n}{N}$, $\mathrm{Var}\,V_i = n\,m\,N^{-2}$ and $\mathrm{Cor}(V_i,V_j) = -nm\,N^{-2}(N-1)^{-1}$ $(i \neq j)$. For a two-sample linear rank statistic these results immediately imply that under the assumption $F = G$

$$(3.2.9) \qquad ET = n\,N^{-1} \sum_{i=1}^{N} a(i)$$

and

$$(3.2.10) \qquad \mathrm{Var}\,T = n\,m(N-1)^{-1}\{N^{-1} \sum_{i=1}^{N} a(i)^2 - \bar{a}^2\}$$

(Exercise: Deduce (3.2.9) and (3.2.10) from Lemma 3.2.1). The scores $a(i) = i$ are called the *Wilcoxon scores*. Commonly used are also the following special scores:

(a) *van der Waerden scores:* $a(i) = \varphi(\frac{i}{N+1})$, where φ denotes the inverse of the standard normal d.f.
$\Phi(t) = \frac{1}{\sqrt{2\pi}} \int_{-\infty}^{t} e^{-1/2\,u^2}\,du$. Because of symmetry, for the corresponding two-sample linear rank statistic T we obtain $ET = 0$, and hence $\mathrm{Var}\,T = 2\,mn\,(N(N-1))^{-1} \sum_{i=1}^{[N/2]} \varphi^2(\frac{i}{N+1})$. This statistic is called the *van der Waerden test statistic* and tests based on it belong to the class of *normal score tests* (where the scores are defined as functions of Φ^{-1}).

(b) *Fisher-Yates-Terry-Hoeffding scores:* Let $Z_{(1)},\ldots,Z_{(N)}$ denote the ordered sample of N i.i.d. $N(0,1)$-distributed random variables and define $a(i) = E\,Z_{(i)}$. Then $ET = 0$ and $\mathrm{Var}\,T = mn(N(N-1))^{-1} \sum_{i=1}^{N} (E\,Z_{(i)})^2$. Note that $E\,Z_{(i)} = E\,\varphi\,U_{(i)}$ where φ is as in (a) and

$\{U_{(i)} : 1 \leq i \leq N\}$ denotes the ordered sample of N i.i.d. uniformly distributed random variables. Thus, the $a(i)$ are also normal scores. Theorem 3.2.1 shows that the Fisher-Yates-Terry-Hoeffding test is locally most powerful if the density f of the d.f. F satisfies $|f'(x)| = |x| f(x)$ and the assumptions (3.2.4) and (3.2.5). For example, $f(x) \sim 1_{[a,b]} e^{-1/2\, x^2}$ is such a density.

All statistics discussed in this example may serve to test for location in a family $F_\vartheta(x) = F(x-\vartheta)$ $(\vartheta \in \mathbb{R})$.

Example 3.2.2:

While in the last examples we introduced some tests for location alternatives, we shall now discuss their analogues for tests of scale alternatives. Consider a family G_τ $(\tau>0)$ of d.f. where

$$G_\tau(x) = F\left(\frac{x-\mu}{\tau}\right)$$

for some unknown d.f. F and some fixed unknown μ. The hypothesis is specified by $\tau = 1$. {If μ is not fixed *('dispersion problem')*, the following tests are also used; however, there are examples where no rank test is good (Exercise).} If F has a density f then G_τ has the density $g_\tau(x) = \tau^{-1} f(\tau^{-1}(x-\mu))$, which can be seen by integration. Furthermore, if $\int x\, dF(x) = 0$,

$$\int x\, dG_\tau(x) = \tau \int x\, dF(x) + \mu = \mu$$

and

$$\int x^2 dG_\tau(x) = \tau^2 \int x^2\, dF(x) + \mu^2.$$

Hence, $\operatorname{Var} G_\tau = \tau^2 \operatorname{Var} F$, and the hypothesis becomes $\operatorname{Var} X = 1$.

(a) The analogue to the Fisher-Yates-Terry-Hoeffding test is known as the *Capon* (normal scores) *test*. The scores are given by $a(i) = E\, Z^2_{(i)}$ where $Z_{(1)} < \ldots < Z_{(N)}$ denote the order statistic of N i.i.d. $\mathcal{N}(0,1)$-distributed random variables (as before). Since $E \sum_{i=1}^{N} Z^2_{(i)} = N$, we get from Lemma 3.1.1 that for $T = \sum_{i=1}^{n} a(r_i)$

$$ET = N^{-1} \sum_{i=1}^{N} c_i \sum_{j=1}^{N} a(j) = n$$

and

$$\text{Var } T = nm(N(N-1))^{-1} \sum_{i=1}^{N} (E\, Z_{(i)}^2)^2 - \frac{n\, m}{N-1} .$$

The optimality property discussed for the Fisher-Yates-Terry-Hoeffding test transfers (Theorem 3.2.1).

(b) The *Klotz test* corresponds to the van der Waerden-test. The scores are defined by

$$a(i) = (\Phi^{-1}(\tfrac{i}{N+1}))^2 \qquad (1 \leq i \leq N),$$

hence

$$E\, T = n\, N^{-1} \sum_{i=1}^{N} (\Phi^{-1}(\tfrac{i}{N+1}))^2$$

and

$$\text{Var } T = nm(N(N-1))^{-1} \sum_{i=1}^{N} (\Phi^{-1}(\tfrac{i}{N+1}))^4 - \frac{m}{n(N-1)} (ET)^2 .$$

(c) The *Ansari-Bradley test* may be considered as the analogue to the Wilcoxon 2-sample test. Here

$$a(i) = \frac{N+1}{2} - \left| i - \frac{N+1}{2} \right| .$$

The expectation and variance of the two-sample linear rank statistic T under $F = G$ are given by (Exercise)

$$ET = \begin{cases} 1/4\; n(N+2) & \text{if } N \text{ is even} \\ 1/4\; n\, N^{-1}(N+1)^2 & \text{if } N \text{ is odd} \end{cases}$$

and

$$\text{Var } T = \begin{cases} (48(N-1))^{-1} nm(N^2-4) & \text{if } N \text{ is even} \\ (24\, N)^{-1}\, nm\, (N^2-9) & \text{if } N \text{ is odd.} \end{cases}$$

(d) The *Siegel-Tukey test* has an even closer connection to the Wilcoxon two-sample test than the previous one because under the hypotheses it has the same distribution as the Wilcoxon test. The scores are defined by

$$a(i) = \begin{cases} 2i & \text{if } i \text{ is even and } 1\le i\le n/2 \\ 2i-1 & \text{if } i \text{ is odd and } 1\le i\le n/2 \\ 2(N-i)+2 & \text{if } N-i \text{ is even and } N/2\le i\le N \\ 2(N-i)+1 & \text{if } N-i \text{ is odd and } N/2\le i\le N. \end{cases}$$

Since a maps $\{1,\ldots,N\}$ onto itself, $ET = \frac{1}{2}\, n(N+1)$ and $\mathrm{Var}\, T = 1/12\, nm(N+1)$ (Exercise).

For this statistic the "Wilcoxon" weights $1,\ldots,N$ are given to the variables $X_{(1)}, X_{(N)}, X_{(N-1)}, X_{(2)}, X_{(3)}), \ldots$. This means that large and small observations are of small influence on the value of T , while the medium values have the large weights. If the variance of the second sample is larger than that of the first one, heuristically, more observations lie further apart from the expected value μ , and T will also be large.

Example 3.2.3: c-sample problems are an immediate generalization of the location problem in Example 3.2.1. Consider c-independent samples $\{X_{i,j} : 1\le i\le n_j\}$ $(1\le j\le c)$ of i.i.d. F_j-distributed random variables. Here the hypotheses is $F_j = F_1$ for all $j=2,\ldots,c$, where it is assumed that $F_j(x) = F(x-\vartheta_j)$. In the parametric case, $F_j = N(\vartheta_j,\sigma^2)$ (σ^2 unknown), the F-test has to be used. The problem has a formulation as a linear model setting

$$X_{ij} = \mu + \alpha_j + \varepsilon_{ij}$$

where the ε_{ij} are i.i.d. F-distributed, and the hypothesis is given by $\alpha_j = \alpha_1$ $(2\le j\le c)$. (W.l.o.g. one always may assume that $F_1 = F$ or $\alpha_1 = 0$.)

The *Kruskal-Wallis test* may be considered as the nonparametric version. Denote by R_{ij} the rank of X_{ij} among all observations $\{X_{k,l} : 1\le k\le n_l, 1\le l\le c\}$, and let $R_j = \sum_{i=1}^{n_j} R_{ij}$.

Under the hypothesis, the R_{ij} are identically distributed, and hence

$$ER_j = \sum_{i=1}^{n_j} ER_{ij} = 2^{-1} n_j (N+1)$$

where $N = \sum_{j=1}^{c} n_j$. The statistic is defined by

$$T = 12\ (N(N+1))^{-1} \sum_{j=1}^{c} n_j^{-1} (R_j - 2^{-1} n_j (N+1))^2,$$

which measures the sum of the squares of the difference of the simple linear rank statistic R_j and its expectation. The asymptotic distribution of T can be computed from that of the simple linear rank statistic. (Exercise: Use Theorem 3.4.3 below.) Other statistics can be formulated analogously to the Fisher-Yates-Terry-Hoeffding and the van der Waerden statistic.

Example 3.2.4: The *Friedman test* for random blocks is similar to the Kruskal-Wallis test. Consider the problem of testing the effect of c treatments. Here n blocks are observed, where each block contains one observation for each treatment. Thus each block j = 1,...,n consists of random variables $X_{1j},...,X_{cj}$ with continous d.f. $F_{ij}(x) = F_j(x-\vartheta)$ and the hypothesis is given by $F_{1j} = F_{2j} = ... = F_{cj}$ for each j. The Friedman test statistic is defined by

$$T = 12\ (nc(c+1))^{-1} \sum_{i=1}^{c} [\sum_{j=1}^{n} R_{ij} - 2^{-1} nc(c+1)]^2$$

where R_{ij} denotes the rank of X_{ij} among all observations. The asymptotic distribution is determined as in Example 3.2.3.

Example 3.2.5: We finally discuss rank tests for independence. Let $(X_1,Y_1),...,(X_N,Y_N)$ be N two-dimensional i.i.d. random variables with continuous one-dimensional marginals. Denote by R_i the rank of X_i among $X_1,...,X_N$ and by Q_i the rank of Y_i among $Y_1,...,Y_N$. Here the hypothesis is the independence of X_1 and Y_1. In the normal case it suffices to test for zero correlation, but here an appropriate statistic is given by $T(x,y) = \sum_{i\leq N} a(r_i)\, a(q_i)$, where r_i (resp. q_i) denotes the rank of x_i (resp. y_i) among $x_1,...,x_N$ (resp. $y_1,...,y_N$). If the scores a(i) are given by $a(i) = EV_i$, where $V_1 < V_2 < ... < V_N$ denotes the ordered sample of N i.i.d. observations from $\mathcal{N}(0,1)$, we have again a Fisher-Yates (normal score) test statistic. Each rank vector $r(x)$ may be written in a unique way as $(\tau(1),\ldots,\tau(N))$ with $\tau \in \gamma_N$. The vector $d(x) = (\tau^{-1}(1),\ldots,\tau^{-1}(N))$ is called the *antirank* of x. If the rank vector $q(y)$ corresponds to $\sigma \in \gamma_N$ then we may write

$$T = \sum_{i=1}^{N} a(\tau(i))\, a(\sigma(i)) = \sum_{i=1}^{N} a(i)\, a(\sigma(\tau^{-1}(i)))$$
$$= \sum_{i=1}^{N} a(i)\, a(r_i^o)$$

where r_i^o denotes the rank of $y_{\tau^{-1}(i)}$ among $y_1,\ldots,y_N$, hence T may be considered as a composition of two rank statistics. Under the hypothesis the distribution of $(Y_1,\ldots,Y_N)$ is the same as that of $(Y_{\tau^{-1}(1)},\ldots,Y_{\tau^{-1}(N)})$ where $\tau(i)=R_i$, the rank of X_i among $X_1,\ldots,X_N$ (Exercise). Therefore we obtain from Lemma 3.2.1 that

$$\text{(3.2.9)} \qquad ET = N\,\bar{a}^2$$

and

$$\text{(3.2.10)} \qquad \operatorname{Var} T = (N-1)^{-1} \sum_{i,j=1}^{N} (a(i)a(j)-\bar{a}a(j)-\bar{a}a(i)+\bar{a}^2)^2$$

For the Fisher-Yates scores $\bar{a} = 0$ and hence $ET = 0$ and $\operatorname{Var} T = (N-1)^{-1}(\sum a(i)^2)^2$. Similar results hold for the van der Waerden scores $a(i) = \Phi^{-1}(i/N+1)$. Using the Wilcoxon scores $a(i) = i$ we obtain the *Spearman correlation coefficient*. $\sum_{i \le N} i\, R_i^o$ is equivalent to the statistic

$$\rho = 12\,(N^3-N)^{-1} \sum_{i=1}^{N} (i-(N+1)/2)\,(R_i^o-(N+1)/2)\ .$$

Here $ET = 4^{-1}N(N+1)^2$ and $\operatorname{Var} T = 144^{-1}\, N^2(N+1)^2(N-1)$, equivalently $E\rho = 0$ and $\operatorname{Var} \rho = (N-1)^{-1}$. (Exercise)

3. A representation of simple linear rank statistics

This section serves as preparation of the following three parts. We represent a simple linear rank statistic as a stochastic integral and investigate the Banach space structure of a certain subclass of score functions, for which we shall prove the asymptotic normality in section 4.

Let $T = \sum_{i=1}^{N} c_i\, a(R_i)$ denote a simple linear rank statistic based on N independent observations $X_1,\ldots,X_N$ with continuous d.f. F_i of X_i $(1 \le i \le N)$. The empirical d.f. of $X_1,\ldots,X_N$ is denoted by

$$(3.3.1) \qquad \hat{H}(t) = N^{-1} \sum_{i=1}^{N} 1_{\{X_i \le t\}}$$

and similarly the empirical d.f. of $X_1,\ldots,X_N$, weighted by c_i, is denoted by

$$(3.3.2) \qquad \hat{F}(t) = N^{-1} \sum_{i=1}^{N} c_i \, 1_{\{X_i \le t\}}$$

(Note that the notation is slightly different to the first two chapters!)

Both, $\hat{H}$ and $\hat{F}$, are right continuous d.f. and we can associate their left continuous inverses by setting

$$(3.3.3) \qquad \begin{aligned} \hat{H}^{-1}(z) &= \inf\{t \in \mathbb{R}: \hat{H}(t) \ge z\} \\ \hat{F}^{-1}(z) &= \inf\{t \in \mathbb{R}: \hat{F}(t) \ge z\}. \end{aligned}$$

The ranks R_i of X_i can be expressed by $\hat{H}$, since

$$R_i = N\,\hat{H}(X_i) \qquad (1 \le i \le N)$$

and hence

$$T = \sum_{i=1}^{N} c_i \, a(N\hat{H}(X_i)),$$

has the following representation:

<u>Lemma 3.3.1:</u> Let $T = \sum_{i \le N} c_i \, a(R_i)$ be a simple linear rank statistic and let $h:(0,1) \longrightarrow \mathbb{R}$ be any function satisfying

$$(3.3.4) \qquad h(i/N+1) = a(i) \qquad (1 \le i \le N).$$

Then T can be represented as a stochastic integral

$$(3.3.5) \qquad T = N \int_{-\infty}^{\infty} h\left(\frac{N}{N+1} \hat{H}(t)\right) d\hat{F}(t).$$

h is called a *score function.*

A special score function (given the scores a(i)) is $h_a(t) = a(i)$ for $i/N+1 \le t < (i+1)/N+1$ and $h_a(t) = 0$ if $t < 1/(N+1)$. Lemma 3.2.1 immediately yields

$$ET = \left(\sum_{i \le N} c_i\right) \frac{N+1}{N} \int h_a(t)\, dt$$

and

$$\operatorname{Var} T = \frac{N+1}{N-1} \sum_{i=1}^{N} (c_i - \bar{c})^2 \left(\int_0^1 h_a^2(t)\,dt - \frac{N+1}{N} \left(\int_0^1 h_a(t)\, dt \right)^2 \right).$$

Note that h_a is right continuous and has bounded variation. Functions of this kind will now be investigated in some detail.

Lemma 3.3.2: Let $h:(0,1)\longrightarrow\mathbb{R}$ be a right continuous function which has bounded total variation on each compact subset of $(0,1)$ and which satisfies $h(1/2) = 0$. Then there exist two σ-finite measures ν_1 and ν_2 on $(0,1)$ with the following properties:

(3.3.6) $\quad \nu_1$ and ν_2 are orthogonal.

(3.3.7) $\quad h(t) = h_1(t) + h_2(t) \quad (0<t<1)$, where

$$h_i(t) = \begin{cases} \nu_i((1/2,t]) & \text{if } t\geq 1/2 \\ \nu_i((t,1/2]) & \text{if } t\leq 1/2 \end{cases} \qquad (i=1,2).$$

This decomposition is unique.

Proof: Define the signed measure ν on $(0,1)$ by

$$\nu((1/2,t]) = h(t) \quad (t\geq 1/2), \quad \nu((t,1/2]) = h(t) \quad (t\leq 1/2).$$

(This definition extends uniquely to a σ-finite signed measure on $(0,1)$.) By the Hahn-Jordan decomposition theorem there exists a unique representation $\nu = \nu_1 - \nu_2$ of ν as a difference of two orthogonal, positive, σ-finite measures ν_1 and ν_2. This proves the lemma.

We shall frequently use partial integration of Lebesgue-Stieltjes integrals. The basic formula is this: Let f and g be two functions of bounded variation on an interval $A=[a,b]$. Then

(3.3.8) $$\int_A f(t+)\, dg(t) + \int_A g(t-)\, df(t) = f(b+)g(b+)-f(a-)g(a-).$$

We also remark the following substitution rule. Let $f:I\longrightarrow J$ and $g:I\longrightarrow\mathbb{R}$ be non-decreasing. If f^{-1} denotes the left continuous inverse of f, then for every $dg\circ f^{-1}$ integrable function φ

(3.3.9) $$\int_{[a,b)} \varphi\, dg\circ f^{-1} = \int_{[f^{-1}(a),f^{-1}(b))} \varphi\circ f\, dg .$$

Lemma 3.3.3: Let $h:(0,1)\longrightarrow\mathbb{R}$ be right continuous, satisfying $h(1/2)=0$ and having bounded variation on each compact subset of $(0,1)$. Define

(3.3.10) $\quad F = E\hat{F} \quad$ and $\quad H = E\hat{H}$.

Then, using the notation of (3.3.5)

$$N^{-1}T - \int_{-\infty}^{\infty} h(H(t))\, dF(t) = -\int_0^1 (\hat{F}(\hat{H}^{-1}(\tfrac{N+1}{N}s))-F(H^{-1}(s)))dh(s)$$

provided $h\circ H \in L_1(F)$.

Proof: By Lemma 3.3.2 it suffices to consider an increasing function h. First, (3.3.9) implies that

$$N^{-1}T = \int h(\tfrac{N}{N+1}\hat{H}(t))\; d\hat{F}(t) =$$

$$= \int 1_{\left[(\frac{N}{N+1}\hat{H})^{-1}(\frac{1}{N+1}),\,(\frac{N}{N+1}\hat{H})^{-1}(\frac{N}{N+1})\right]}(t)\, h(\tfrac{N}{N+1}\hat{H}(t))\; d\hat{F}(t)$$

$$= \int 1_{[\frac{1}{N+1},\frac{N}{N+1})}(s)\; h(s)\; d\hat{F}\circ(\tfrac{N}{N+1}\hat{H})^{-1}(s) +$$

$$+ h(\tfrac{N}{N+1})\;(\hat{F}(\hat{H}^{-1}(1)) - \hat{F}(\hat{H}^{-1}(1)-)) =$$

$$= \int_0^1 h(t)\; d\hat{F}\circ(\tfrac{N}{N+1}\hat{H})^{-1}(t).$$

and

$$\int_{-\infty}^{\infty} h(H(t))\; dF(t) = \int_0^1 h(t)\; dF\circ H^{-1}(t)\;.$$

Now apply (3.3.8) to obtain

$$\int_a^{1/2} h(t)\; d\hat{F}\circ(\tfrac{N}{N+1}\hat{H})^{-1}(t) = -\int_a^{1/2} \hat{F}\circ(\tfrac{N}{N+1}\hat{H})^{-1}(t)\; dh(t)$$

(a small),

$$\int_{1/2}^{b} h(t)\; d\hat{F}\circ(\tfrac{N}{N+1}\hat{H})^{-1}(t) = \int_{1/2}^{b} (1 - \hat{F}\circ(\tfrac{N}{N+1}\hat{H})^{-1}(t))\, dh(t)$$

$$= h(b) - \int_{1/2}^{b} \hat{F}\circ(\tfrac{N}{N+1}\hat{H})^{-1}(t)\; dh(t) \quad \text{(b large)},$$

$$\int_a^{1/2} h(t)\; dF\circ H^{-1}(t) = -h(a-)F\circ H^{-1}(a-) - \int_a^{1/2} F\circ H^{-1}(t)\; dh(t)$$

and

$$\int_{1/2}^{b} h(t)\; dF\circ H^{-1}(t) = -h(b)(1-F\circ H^{-1}(b)) +$$

$$\int_{1/2}^{b} (1-F\circ H^{-1}(t))\; dh(t) =$$

$$= h(b)\; F\circ H^{-1}(b) - \int_{1/2}^{b} F\circ H^{-1}(t)\; dh(t).$$

If $a \longrightarrow 0$ and $b \longrightarrow 1$, $h(a)\; F\circ H^{-1}(a) \longrightarrow 0$ and $h(b)(1-F\circ H^{-1}(b)) \longrightarrow 0$. Putting everything together proves the lemma.

We shall not make further use of Lemma 3.3.3. However, if there is some need to extend the asymptotic result below to other than absolutely continuous score functions, the representation in 3.3.3 can easily be used to prove asymptotic normality in the case where dh is a finite signed measure only.

The extension method in section 4 works for general h.

<u>Definition 3.3.1:</u> A score function $h:(0,1)\longrightarrow \mathbb{R}$ is called *proper*, if it is right continuous, has bounded total variation on each compact subset of $(0,1)$, satisfies $h(1/2) = 0$ and has finite norm $\|h\| < \infty$, where $\|h\|$ is defined by

$$(3.3.11) \qquad \|h\| = \int_0^1 (z(1-z))^{-1/2} (|h_1(z)| + |h_2(z)|)\, dz$$

using the decomposition $h = h_1 - h_2$ of Lemma 3.3.2. We denote by $\mathcal{H}$ the set of all proper score functions.

We first need

<u>Lemma 3.3.4:</u> Let h be a right continuous score function with bounded total variation on each compact subset of $(0,1)$ satisfying $h(1/2) = 0$. Then

$$(3.3.12) \quad \|h\| < \infty \Rightarrow \quad \limsup_{z\to\{0,1\}} |h(z)|(z(1-z))^{1/2} = 0$$

and

$$\sup_{0<z<1} |h(z)|(z(1-z))^{1/2} \leq \|h\| .$$

$$(3.3.13) \quad \int_0^1 (z(1-z))^{1/2} |dh|(z) \leq \|h\| \leq 8 \int_0^1 (z(1-z))^{1/2} |dh|(z).$$

where $|dh|$ denotes the total variation measure of dh.

$$(3.3.14) \quad \int_0^1 h(z)^2\, dz \leq \|h\|^2.$$

<u>Proof:</u> By Lemma 3.2.2 it suffices to prove the lemma for an increasing function h. (The definition of the norm in (3.3.11) is compatible with the decomposition of h.) We also may assume that $h(z) \equiv 0$ for $z \leq 1/2$.

First note that

$$(3.3.15) \qquad d:= \liminf_{z\longrightarrow 1} h(z)\ (z(1-z))^{1/2} = 0$$

if $\|h\| < \infty$. Otherwise one could find some z_0 with $h(z)(z(1-z))^{1/2} \geq 2^{-1}d$ for $z_0 \leq z$, contradicting

$$\infty > \|h\| \geq \int_{z_0}^1 h(z)(z(1-z))^{-1/2}dz \geq$$

$$\geq 2^{-1}d \int_{z_0}^1 (z(1-z))^{-1}dz.$$

Using partial integration we obtain for any $b < 1$ that

$$(3.3.16)\quad 2^{-1}\int_0^b h(z)\;(z(1-z))^{-1/2}(1-2z)\;dz = h(b)(b(1-b))^{1/2} - \int_0^b (z(1-z))^{1/2}dh(z).$$

Let us show (3.3.13).

If $\|h\| = \infty$, then $\int (z(1-z))^{1/2}\;dh(z) = \infty$, since otherwise the right-hand side of (3.3.16) is bounded from below, while the left-hand side approaches $-\infty$.

If $\|h\| < \infty$, then from (3.3.15) and (3.3.16) we obtain

$$2^{-1}\int_0^1 h(z)\;(1-2z)(z(1-z))^{-1/2}dz = -\int_0^1 (z(1-z))^{1/2}dh(z).$$

Since for $z \geqq 2^{-1}$, $0 \geqq 1-2z \geqq -1$ we have

$$\int_0^1 (z(1-z))^{1/2}\;dh(z) \leq \|h\|.$$

The monotonicity of $h(z)\;(z(1-z))^{-1/2}$ for $z \geqq 2^{-1}$ implies that

$$\|h\| = \int_{1/2}^1 h(z)\;(z(1-z))^{-1/2}\;dz \leq 2\int_{3/4}^1 h(z)(z(1-z))^{-1/2}dz \leq$$

$$\leqq 4\int_{3/4}^1 h(z)\;(2z-1)\;(z(1-z))^{-1/2}dz$$

$$\leqq 8\int_0^1 (z(1-z))^{1/2}\;dh(z).$$

We proceed with (3.3.12).

Let $\varepsilon > 0$ be given. The last equality shows that there exists some b_0 such that for $b \geqq b_0$

$$\left|2^{-1}\int_0^b h(z)\;(2z-1)\;(z(1-z))^{-1/2}\;dz - \int_0^b (z(1-z))^{1/2}dh(z)\right| < \varepsilon.$$

But this difference is $h(b)(b(1-b))^{1/2}$ by (3.3.16), and the first relation in (3.3.12) is proven. The second inequality is obtained from

$$h(b)\;(b(1-b))^{1/2} = \int_0^b (b(1-b))^{1/2}dh(z) \leq$$

$$\leqq \int_0^b (z(1-z))^{1/2}\;dh(z) \leqq \|h\|.$$

It is left to show (3.3.14), which now follows immediately

from (3.3.12) and

$$\int h(z)^2\,dz = \int h(z)(z(1-z))^{1/2}\,h(z)(z(1-z))^{-1/2}\,dz \leq$$
$$\leq \|h\|^2.$$

Example 3.3.1: Let $h = G^{-1}$ be the inverse of a continuous d.f. G. Then

$$\|G^{-1}\| = \int_{-\infty}^{\infty} |u|(G(u)(1-G(u)))^{-1/2}\,dG(u)\,.$$

Especially, if $G = \Phi$ the d.f. of $\mathfrak{N}(0,1)$, then for large u we have

$$1 - \Phi(u) \in (2\pi)^{-1}\exp(-u^2/2)\;[(u(1+u^{-2}))^{-1}\,,\,u^{-1}]$$

and hence

$$\|\Phi^{-1}\| \leq \int_{-\infty}^{\infty} p(u)\exp(-u^2/4)\,du < \infty$$

where p is some polynomial.

Lemma 3.3.5: $(\mathfrak{H}, \|\ \|)$ is a Banach space. The set $\mathfrak{G}$ of all absolutely continuous proper score functions forms a closed subspace, and the set $C_{2,b}$ of all functions on (0,1) with bounded second derivative is dense in $\mathfrak{G}$.

Proof: For $h \in \mathfrak{H}$ denote by ν the corresponding σ-finite signed measure on (0,1). Their decompositions according to Lemma 3.3.2 are denoted by $h = h_1 - h_2$ and $\nu = \nu_1 - \nu_2$ with $dh_i = d\nu_i$ (i=1,2). The total variation norm $|\mu|_T$ of a finite signed measure μ on (0,1) satisfies

$$|\mu|_T = \mu_1((0,1)) + \mu_2((0,1))$$

where $\mu = \mu_1 - \mu_2$ denotes the Hahn-Jordan decomposition of μ. Define $\tilde{\nu}$, $\tilde{\nu}_1$ and $\tilde{\nu}_2$ by their density $(z(1-z))^{1/2}$ with respect to ν, ν_1 and ν_2 respectively. Then $\tilde{\nu}=\tilde{\nu}_1 - \tilde{\nu}_2$ is the Hahn-Jordan decomposition of $\tilde{\nu}$ and hence

$$|\tilde{\nu}|_T = \int (z(1-z))^{1/2}(dh_1(z)+dh_2(z)).$$

From (3.3.13) we conclude that

$$(1/8)\,\|h\| \leq |\tilde{\nu}|_T \leq \|h\|$$

and therefore the map $h \longrightarrow \tilde{\nu}$ is an injective, bounded linear map. It follows that $\mathfrak{H}$ is a Banach space, because the finite signed measures with the total variation norm form a Banach space.

Clearly, absolutely continuous $\tilde{\nu}$ correspond to absolutely continuous h, and therefore $\mathbb{G}$ is a closed subspace. In order to see that $\mathbb{G}$ is the norm closure of $C_{2,b}$, fix $h \in \mathbb{G}$. Then h_1 and h_2 belong to $\mathbb{G}$ and $dh_i(z) = h_i'(z)dz$. Hence the claim follows, since we may approximate h_i' by functions in $C_{1,b}$ with respect to $L_1((z(1-z))^{1/2}dz)$, and since by (3.3.13) this is equivalent to approximating h by functions in $C_{2,b}$ with respect to the norm $\| \ \|$.

We conclude this section with a very useful lemma for the computation of random integrals involving empirical processes.

<u>Lemma 3.3.6:</u> For every $m \in \mathbb{N}$ there exists a constant $c = c(m)$ with the following property: Let $g \in L_2(\lambda)$ and $N \in \mathbb{N}$. Assume that $X_1,\ldots,X_N$ are independent random variables with distributions $F_1,\ldots,F_N$ and that $C_1,\ldots,C_N$ are constants satisfying $|C_i| \leq 1$ $(1 \leq i \leq N)$. Then (with the notation in (3.3.1), (3.3.2) and (3.3.10))

$$\text{(3.3.17)} \qquad E\left(\int_{-\infty}^{\infty} g(H(t))\, \left|\frac{N}{N+1}\hat{H}(t) - H(t)\right|^m d\hat{H}(t)\right)^2 \leq$$

$$\leq c \sum_{l=0}^{m} N^{-2m+l} \int g^2(t)\, (t(1-t))^l\, dt$$

and

$$\text{(3.3.18)} \qquad E\left(\int_{-\infty}^{\infty} g(H(t))\, \left(\frac{N}{N+1}\hat{H}(t) - H(t)\right) d(\hat{F}-F)(t)\right)^2 \leq$$

$$\leq c\, N^{-2} \int g^2(t)\, dt \ .$$

<u>Proof:</u> Let us first prove (3.3.17). Note that

$$E(\hat{H}(t) - H(t))^{2m} = \sum_{i_1=1}^{N} \cdots \sum_{i_{2m}=1}^{N} N^{-2m} E \prod_{j=1}^{2m} (1_{\{X_{i_j} \leq t\}} - F_{i_j}(t))$$

$$= \sum_{l=1}^{m} \sum_{1 \leq i_1 < \ldots < i_l \leq N} N^{-2m} \sum_{\substack{\alpha_j \geq 2,\, j=1,\ldots,l \\ \alpha_1+\ldots+\alpha_l = 2m}} \frac{(2m)!}{\alpha_1! \ldots \alpha_l!} \,\cdot$$

$$\cdot\, E \prod_{k=1}^{l} (1_{\{X_{i_k} \leq t\}} - F_{i_k}(t))^{\alpha_k}$$

since $E \prod_{j=1}^{2m} (1_{\{X_{i_j} \leq t\}} - F_{i_j}(t)) = 0$ if some i_j is different from all the other indices.

$\sum_{\alpha_1,\ldots,\alpha_l} \frac{(2m)!}{\alpha_1!\ldots\alpha_l!}$ is the number of partitions of $1,\ldots,2m$ into l sets of cardinality ≥ 2 and hence bounded by some constant depending on m only.

Next observe that

$$\left(\sum_{i=1}^{N} F_i(t)(1-F_i(t))\right)^l = \sum_{i_1,\ldots,i_l=1}^{N} \prod_{k=1}^{l} F_{i_k}(t)(1-F_{i_k}(t))$$

$$\geq \sum_{1\leq i_1<\ldots<i_l\leq N} \prod_{k=1}^{l} F_{i_k}(t)(1 - F_{i_k}(t))$$

and that by Hölder's inequality

$$\text{(3.3.19)} \qquad N^{-1} \sum_{i=1}^{N} F_i(t)\ (1 - F_i(t)) =$$

$$= H(t)(1 - H(t)) + N^{-2} \sum_{i,j=1}^{N} F_i(t)F_j(t) - N^{-1} \sum_{i=1}^{N} F_i^2(t) \leq$$

$$\leq H(t)(1 - H(t)).$$

Putting everything together one obtains

$$E(\hat{H}(t) - H(t))^{2m} \leq$$

$$\sum_{l=1}^{m} \sum_{1\leq i_1<\ldots<i_l\leq N} N^{-2m} \sum_{\alpha_j} \frac{(2m)!}{\alpha_1!\ldots\alpha_l!} \prod_{k=1}^{l} F_{i_k}(t)(1-F_{i_k}(t))$$

$$\leq c_o \sum_{l=1}^{m} N^{-2m+l} (H(t)(1 - H(t)))^l.$$

Integrating $(\hat{H}(X_i) - H(X_i))^{2m}$ with respect to X_j $(j\neq i)$ and using the previous inequality for $t=X_i$ and $(a+b)^{2m} \leq 2^{2m}\cdot(a^{2m} + b^{2m})$ it follows that

$$E(\hat{H}(X_i) - H(X_i))^{2m} \leq 2^{2m}(N^{-2m} + c_o \sum_{l=1}^{m} N^{-2m+l}(H(X_i)(1-H(X_i))^l).$$

Using Hölder's inequality, (3.3.17) is a consequence of

$$E\left(\int g(H(t))\ |\hat{H}(t) - H(t)|^m\ d\hat{H}(t)\right)^2 =$$

$$= E\left(\sum_{1\leq i\leq N} N^{-1}\ g(H(X_i))\ |\hat{H}(X_i) - H(X_i)|^m\right)^2 \leq$$

$$\leq N^{-1} \sum_{1\leq i\leq N} Eg^2(H(X_i))(\hat{H}(X_i) - H(X_i))^{2m} \leq$$

$$\leq N^{-1} \sum_{1\leq i\leq N} E\Big\{ g^2(H(X_i))2^{2m}(N^{-2m}+c_o \sum_{l=1}^{m} N^{-2m+l}$$

$$(H(X_i)(1-H(X_i))^l)\Big\} \leq$$

$$\leq c \sum_{l=0}^{m} N^{-2m+l-1} \sum_{i=1}^{N} E\Big(g^2(H(X_i)(H(X_i)(1-H(X_i)))^l\Big) =$$

$$= c \sum_{l=0}^{m} N^{-2m+l-1} \sum_{i=1}^{N} \int_{-\infty}^{\infty} g^2(H(t))(H(t)(1-H(t)))^l \, dF_i(t) =$$

$$= c \sum_{l=0}^{m} N^{-2m+l} \int_0^1 g^2(t)(t(1-t))^l \, dt.$$

We still have to prove (3.3.18). Define for $1 \leq i,j \leq N$

$$U(i,j) = g(H(X_i))1_{\{X_j\leq X_i\}} - g(H(X_i))F_j(X_i) -$$

$$- \int g(H(t)) \, 1_{\{X_j\leq t\}} \, dF_i(t) + \int g(H(t))F_j(t) \, dF_i(t)$$

and write

$$E\Big(\int g(H(t))(\tfrac{N}{N+1}\hat{H}(t) - H(t)) \, d(\hat{F}-F)(t) \Big)^2 =$$

$$= E\Big(\sum_{1\leq i,j\leq N} N^{-2}c_iU(i,j) - (N+1)^{-1} \int g(H(t))\hat{H}(t) \, d(\hat{F}-F)(t) \Big)^2$$

$$\leq 4 \, E\Big(\sum_{1\leq i,j\leq N} N^{-2}c_iU(i,j) \Big)^2 + 4N^{-2} \int g^2(t) \, dt,$$

since

$$(N+1)^{-2} \, E(\int g(H(t))\hat{H}(t) \, d(\hat{F}-F)(t))^2 \leq$$

$$\leq (N+1)^{-2}N^{-1} \sum_{i=1}^{N} c_i \quad E(g(H(X_i))\hat{H}(X_i) - \int g(H(t))\hat{H}(t) \, dF_i(t))^2.$$

Consider four indices $1 \leq i,j,k,l \leq N$. If one of these indices is different from all the others, then clearly

$$E \, U(i,j)U(k,l) = 0.$$

Consequently

$$E\Big(\sum_{1\leq i,j\leq N} U(i,j)N^{-2}c_i \Big)^2 \leq N^{-4} \sum{}^{*} E|U(i,j)U(k,l)|$$

where $\sum^*$ denotes the summation over all indices $1\leq i,j,k,l\leq N$ not satisfying the above condition.

It is not hard to verify (exercise) that for some constant c_o

$$E|U(i,j)U(k,l)| \leq c_o\Big(\int g^2(H(t))\, dF_p(t) + \int g^2(H(t))\, dF_q(t)\Big)$$

if p and q are the (at most) two different indices. Hence for some constant c_1

$$\Sigma^* E|U(i,j)U(k,l)| \leq c_1 N \sum_{i=1}^{N} \int g^2(H(t))\, dF_i(t)$$

and

$$E\Big(\sum_{i,j=1}^{N} N^{-1} c_i U(i,j) \Big)^2 \leq c_1 N^{-2} \int g^2(H(t))\, dH(t).$$

This proves (3.3.18).

4. Asymptotic normality of simple linear rank statistics

Theorem 3.1.1 immediately implies the asymptotic normality of simple linear rank statistics under the hypothesis, if the score function is square integrable.

Theorem 3.4.1: Let $T_N = \sum_{1 \leq i \leq N} c_{iN}\, a_N(R_i)$ $(N \geq 1)$ be a sequence of simple linear rank statistics. Assume that

$$\lim_{N\to\infty} \Big[\sum_{i=1}^{N} (c_{iN} - \bar{c}_N)^2 \Big]^{-1} \max_{1 \leq j \leq N} (c_{jN} - \bar{c}_N)^2 = 0 \tag{3.4.1}$$

and that there exist a constant D>0 and a square integrable monotone function φ on (0,1) such that

$$\begin{aligned} D &\leq N^{-1} \sum_{i=1}^{N} (a_N(i) - \bar{a}_N)^2 \qquad (N \geq 1) \\ |a_N(i) - \bar{a}_N| &\leq |\varphi(i/N+1)| \qquad (N \geq 1,\ 1 \leq i \leq N). \end{aligned} \tag{3.4.2}$$

Then, under the hypothesis (i.e. for each N, the random variables $X_{1N},\ldots,X_{NN}$ are i.i.d.),

$$(\operatorname{Var} T_N)^{-1/2}\, (T_N - ET_N)$$

converges weakly to the standard normal distribution $\mathfrak{N}(0,1)$.

Proof: For simplicity we shall omit the index N. Define

$$\tilde{a}(i) = N^{1/2}(a(i)-\bar{a})\big(\sum (a(j)-\bar{a})^2\big)^{-1/2}$$

and

$$\tilde{c}_i = N^{1/2}(c_i-\bar{c})\big(\sum (c_j-\bar{c})^2\big)^{-1/2}.$$

From Theorem 3.1.1 we conclude that $N^{-1/2}\tilde{T}$ converges weakly to $\mathfrak{N}(0,1)$, where $\tilde{T} = \sum \tilde{c}_i\, \tilde{a}(R_i)$, provided that for every $\varepsilon > 0$

$$\lim_{N\to\infty} \sum_{\substack{i,j=1 \\ |\tilde{a}(i)\tilde{c}_j| > \varepsilon N^{1/2}}} N^{-2}\, \tilde{a}(i)^2 \tilde{c}_j^{\,2} = 0 . \tag{3.4.3}$$

Since $\tilde{T} = N(\sum(a(i)-\bar{a})^2)^{-1/2}(\sum(C_j-\bar{C})^2)^{-1/2}(T-N\bar{C}\bar{a})$, and since by Lemma 3.1.1 $ET = N\bar{C}\bar{a}$ and $\operatorname{Var} T = (N-1)^{-1}\sum(a(i)-\bar{a})^2\sum(C_j-\bar{C})^2$, it follows that $(\operatorname{Var} T)^{-1/2}(T - ET)$ converges to $\mathbf{N}(0,1)$, once (3.4.3) is verified.

Condition (3.4.1) ensures a sequence $K(N)\longrightarrow\infty$ (as $N\longrightarrow\infty$) such that uniformly in $j = 1,\dots,N$

$$|\tilde{a}(i)\tilde{C}_j| \geqq \varepsilon N^{1/2} \quad\Rightarrow\quad |\tilde{a}(i)| \geqq \varepsilon K(N) \qquad (\varepsilon > 0).$$

Therefore it follows from (3.4.2) that

$$\sum_{|\tilde{a}(i)\tilde{C}_j|\geqq \varepsilon N^{1/2}} N^{-2}\tilde{a}(i)^2\tilde{C}_j^2 \leqq \sum_{|\tilde{a}(i)|\geqq\varepsilon K(N)} N^{-1}\tilde{a}(i)^2$$

$$\leqq N^{-1}D^{-1}\sum_{|\varphi(i/N+1)|\geqq\varepsilon K(N)\sqrt{D}}\varphi(i/N+1)^2$$

$$\leqq \text{const. } D^{-1}\int_{\{|\varphi|\geqq\varepsilon K(N)\sqrt{D}\}}\varphi(t)^2\,dt \longrightarrow 0 \quad (\text{as } N\to\infty).$$

In order to find the asymptotic distribution of simple linear rank statistics under alternatives we need the following approximation theorem. Different forms will be used for signed rank statistics in section 5 and linear combinations of a function of the order statistic in section 6.

Theorem 3.4.2: There exists a constant $K>0$ with the following property: Let h be a proper score function, let $C_1,\dots,C_N$ be regression constants satisfying $\max|C_i| \leqq 1$ and let $X_1,\dots,X_N$ be independent random variables with continuous d.f. $F_1,\dots,F_N$. Then, using the notation in (3.3.1), (3.3.2) and (3.3.10), one has

$$(3.4.4)\qquad N\,E\Big\{\int_{-\infty}^{\infty} h(\tfrac{N}{N+1}\hat{H}(t))\,d\hat{F}(t) - \int_{-\infty}^{\infty} h(\tfrac{N}{N+1}H(t))\,d\hat{F}(t)\Big\}^2 \leqq K\,\|h\|^2,$$

$$(3.4.5)\qquad N\,E\Big\{\int_{-\infty}^{\infty} h(\tfrac{N}{N+1}H(t))\,d(\hat{F} - F)(t)\Big\}^2 \leqq K\,\|h\|^2,$$

$$(3.4.6)\qquad N\Big\{\int_{-\infty}^{\infty}(h(H(t)) - h(\tfrac{N}{N+1}H(t)))\,dF(t)\Big\}^2 \leqq K\,\|h\|^2$$

and

$$(3.4.7)\qquad N\,E\Big\{N^{-1}T - \int h(H(t))\,dF(t)\Big\}^2 \leqq K\,\|h\|^2,$$

where T denotes the simple linear rank statistic defined in (3.3.5).

Remark: Here and in the following the bound 1 for the absolute values of the regression constants stands for mathematical convenience. Multiplication by a constant provides the same result for uniformly bounded regression constants.

Proof: Let X_i, C_i $(1 \leq i \leq N)$ and $h \in \mathcal{H}$ be given as in the theorem. By definition of the norm $\| \ \|$ it suffices to consider an increasing proper score function h.

Proof of (3.4.4):

Define

$$A = \{(z,t) : \frac{N}{N+1} H(t) < z \leq \frac{N}{N+1} \hat{H}(t)\} =$$

$$= \{(z,t) : \hat{H}^{-1}(\frac{N+1}{N} z) \leq t < H^{-1}(\frac{N+1}{N} z)\}$$

and

$$B = \{(z,t) : \frac{N}{N+1} \hat{H}(t) < z \leq \frac{N}{N+1} H(t)\} =$$

$$= \{(z,t) : H^{-1}(\frac{N+1}{N} z) \leq t < \hat{H}^{-1}(\frac{N+1}{N} z)\} .$$

Setting

$$\varphi(z,t) = \begin{cases} 1 & \text{if } (z,t) \in A \\ -1 & \text{if } (z,t) \in B \\ 0 & \text{otherwise} \end{cases}$$

we first observe that

$$\left| \int_{-\infty}^{\infty} \varphi(z,t)\, d\hat{F}(t) \right| = N^{-1} \sum_{i=1}^{N} |\varphi(z,X_i)\, C_i| \leq$$

$$\leq \int_{-\infty}^{\infty} |\varphi(z,t)|\, d\hat{H}(t) \leq |\hat{H}(\hat{H}^{-1}(\frac{N+1}{N} z)-) - \hat{H}(H^{-1}(\frac{N+1}{N} z)-)| .$$

Also it is easy to see that

$$|\hat{H}(\hat{H}^{-1}(\frac{N+1}{N} z)-) - \frac{N+1}{N} z| \leq 2 \min (N^{-1}, z) .$$

Using both estimates we conclude that

$$E \left| \int_{-\infty}^{\infty} \varphi(z,t)\, d\hat{F}(t) \right|^2 \leq$$

$$\leq E\Big(2\min(N^{-1}, z) + |\hat{H}(H^{-1}(\frac{N+1}{N} z)-) - \frac{N+1}{N} z| \Big)^2 \leq$$

$$\leq 8 \min (N^{-2}, z^2) + 2\, E\, |\hat{H}(H^{-1}(\frac{N+1}{N} z)-) - \frac{N+1}{N} z|^2 \leq$$

$$\leq 8 \min (N^{-2}, z^2) + 8\, N^{-1} z(1-z).$$

Finally (3.4.4) follows using Cauchy-Schwarz' inequality, Fubini's theorem and (3.3.13), since

$$N\ E\left(\int_{-\infty}^{\infty} (h(\tfrac{N}{N+1}\hat{H}(t) - h(\tfrac{N}{N+1}H(t))\ d\hat{F}(t) \right)^2 =$$

$$= N\ E\left(\int_{-\infty}^{\infty} \int_{0}^{N/N+1} \varphi(z,t)\ d\hat{F}(t)\ dh(z) \right)^2 =$$

$$= N\int_{0}^{N/N+1} \int_{0}^{N/N+1} E\int_{-\infty}^{\infty} \int_{-\infty}^{\infty} \varphi(z,t)\varphi(y,s)\ d\hat{F}(t)d\hat{F}(s)dh(z)\ dh(y) \leq$$

$$\leq N \left(\int_{0}^{N/N+1} (E(\int_{-\infty}^{\infty} \varphi(z,t)\ d\hat{F}(t))^2)^{1/2}\ dh(z) \right)^2 \leq$$

$$\leq \left(8\int_{0}^{N/N+1} \min\ (N^{-1/2}, N^{1/2}z)\ dh(z) + 8\int_{0}^{N/N+1} (z(1-z))^{1/2}\ dh(z) \right)^2 \leq$$

$$\leq K\ \|h\|^2,$$

where we also used

$$\min\ (N^{-1/2}, N^{1/2}z) \leq \text{const.}\ (z(1-z))^{1/2} \quad \text{for } z \leq \frac{N}{N+1}$$

Proof of (3.4.5):

$$N\ E\left(\int_{-\infty}^{\infty} h(\tfrac{N}{N+1}H(t))\ d(\hat{F} - F)(t) \right)^2 =$$

$$= N\ E\left(N^{-1} \sum_{i=1}^{N} c_i\ (h(\tfrac{N}{N+1}H(X_i)) - \int h(\tfrac{N}{N+1}H(t))dF_i(t)) \right)^2 =$$

$$= N^{-1} \sum_{i=1}^{N} c_i^2\ \text{Var}\ (h(\tfrac{N}{N+1}H(X_i))) \leq$$

$$\leq \text{const.} \int_{-\infty}^{\infty} h^2(\tfrac{N}{N+1}H(t))\ d\tfrac{N}{N+1}H(t) =$$

$$= \text{const.} \int h^2(t)\ dt \leq K\ \|h\|^2$$

by (3.3.14).

Proof of (3.4.6):

This follows immediately from

$$N^{1/2} \int_{-\infty}^{\infty} (h(H(t)) - h(\tfrac{N}{N+1}H(t)))\ dF(t) =$$

$$= N^{1/2} \int_{-\infty}^{\infty} \int_{0}^{1} 1_{(\frac{N}{N+1}H(t),H(t)]}(z)\ dh(z)\ dF(t) =$$

$$= N^{1/2} \int_{0}^{1} \int_{-\infty}^{\infty} 1_{[H^{-1}(z),H^{-1}(\frac{N+1}{N}z\wedge 1))}(t)\ dF(t)\ dh(z) \leq$$

$$\leq N^{1/2} \int_{0}^{N/N+1} N^{-1} z\ dh(z) + N^{1/2} \int_{N/N+1}^{1} (1-z)\ dh(z) \leq$$

$$\leq K^{1/2} \| h \| .$$

The proof of (3.4.7) follows from the triangle inequality and (3.4.4) - (3.4.6).

From now on we shall indicate the dependence on N by adding the subscript N to C_i, F, $\hat{F}$, H, $\hat{H}$ etc. By Lemma 3.3.1 a simple linear rank statistic T has a representation $T = \int h(\frac{N}{N+1} \hat{H}_N(t))\ d\hat{F}_N(t)$, where we assume that the score function h is fixed. We define a new statistic $T_N(h)$, which is more convenient to us, by

$$(3.4.8) \quad T_N(h) = N^{1/2}\left\{ \int_{-\infty}^{\infty} h(\tfrac{N}{N+1}\hat{H}_N(t))\ d\hat{F}_N(t) - \right.$$

$$\left. - \int_{-\infty}^{\infty} h(H_N(t))\ dF_N(t) \right\} .$$

Thus (3.4.7) states that

$$\sup_N \ E\ T_N(h)^2 \leq K \|h\|^2$$

under the assumptions in Theorem 3.4.2.

<u>Remark:</u> Define for $h \in \mathcal{H}$

$$(3.4.9) \quad W_N = W_N(h) = N^{1/2}\left\{ \int_{-\infty}^{\infty} \left(h(H_N(t)\ d(\hat{F}_N - F_N)(t) + \right.\right.$$

$$\left.\left. + \int_{-\infty}^{\infty} (\hat{H}_N(t) - H_N(t))\ \frac{dF_N}{dH_N}(t)\ dh(H_N(t)) \right)\right\}$$

and denote by $\sigma_N^2 = \sigma_N^2(h)$ its variance. We have $EW_N=0$ and hence using the partial integration formula (3.3.8) it follows that

$$\sigma_N^2 = N\ E\left(\int_{-\infty}^{\infty} [(\hat{H}_N(t)-H_N(t))\frac{dF_N}{dH_N}(t) - (\hat{F}_N(t-)-F_N(t))]\ dh(H_N(t)) \right)^2$$

since for all sufficiently large b

$$h(H_N(b))(\hat{F}_N(b)-F_N(b)) = h(H_N(b))(1-F_N(b))$$
$$\leq h(H_N(b))(1-H_N(b)) \longrightarrow 0 \quad \text{as } b\longrightarrow\infty$$

by (3.3.12), and similarly for $b\longrightarrow-\infty$.

Let us assume that h is continuous. Then

$$\sigma_N^2 = N\,E\int_{-\infty}^{\infty}\int_{-\infty}^{\infty} [(\hat{H}_N(t)-H_N(t))\frac{dF_N}{dH_N}(t) -(\hat{F}_N(t)-F_N(t))]$$
$$[(\hat{H}_N(s)-H_N(s))\frac{dF_N}{dH_N}(s) - (\hat{F}_N(s)-F_N(s))]\; dh(H_N(t))dh(H_N(s))$$

$$(3.4.10) \quad = 2N^{-1}\sum_{i=1}^{N}\iint_{s<t} F_{iN}(s)(1-F_{iN}(t))(\frac{dF_N}{dH_N}(t)-C_{iN})$$
$$(\frac{dF_N}{dH_N}(s)-C_{iN})\; dh(H_N(s))dh(H_N(t)) .$$

In case that $C_{iN} = 1$ for $i \leq n < N$ and $C_{iN} = 0$ for $i > n$, this reduces to

$$\sigma_N^2 = 2N^{-1}\sum_{i=1}^{n}\iint_{s<t} F_{iN}(s)(1-F_{iN}(t))\,\frac{d(H_N-F_N)}{dH_N}(t)\frac{d(H_N-F_N)}{dH_N}(s)$$
$$dh(H_N(s))dh(H_N(t))$$
$$+\;2N^{-1}\sum_{i=n+1}^{N}\iint_{s<t} F_{iN}(s)(1-F_{iN}(t))\,\frac{dF_N}{dH_N}(s)\,\frac{dF_N}{dH_N}(t)$$
$$dh(H_N(s))dh(H_N(t)).$$

For a two sample linear rank statistic (i.e. $F_{iN}=F$ if $i\leq n$ and $F_{iN}=G$ if $i>n$) one obtains furthermore

$$\sigma_N^2=2\frac{n(N-n)^2}{N^3}\iint_{s<t} F(s)(1-F(t))\,\frac{dG}{dH_N}(t)\,\frac{dG}{dH_N}(s)dh(H_N(s))dh(H_N(t))$$
$$+\;2\,\frac{n^2(N-n)}{N^3}\iint_{s<t} G(s)(1-G(t))\,\frac{dF}{dH_N}(t)\,\frac{dF}{dH_N}(s)\; dh(H_N(s))dh(H_N(t))$$

Moreover, if $nN^{-1}\longrightarrow\lambda_o \in (0,1)$ and if $N\rightarrow\infty$ then

$$\sigma^2(h) = \lim_{N\rightarrow\infty}\sigma_N^2(h) = 2\lambda_o(1-\lambda_o)^2\iint_{s<t} F(s)(1-F(t))\,\frac{dG}{dH}(t)\frac{dG}{dH}(s)$$
$$dh(H(s))dh(H(t))$$
$$(3.4.11)$$
$$+\;2\lambda_o^2(1-\lambda_o)\iint_{s<t} G(s)(1-G(t))\frac{dF}{dH}(t)\frac{dF}{dH}(s)dh(H(s))dh(H(t))$$

where $H(t) = \lambda_o F(t) + (1-\lambda_o)G(t) \quad (-\infty < t < \infty)$.

Definition 3.4.1: A sequence $\{P_n: n \geq 1\}$ of probability measures is said to converge weakly in L_2 to the distribution P if

$$(3.4.12) \qquad \lim_{n\to\infty} D_2(P_n, P) = 0$$

where

$$(3.4.13) \qquad D_2(Q_1,Q_2) = \inf\{ (E(Y_1-Y_2)^2)^{1/2} : Q_i = \mathcal{L}(Y_i), i=1,2\}.$$

We remark that D_2 is in fact a metric (cf. Cambanis et al. 1976). Moreover, if $D_2(P_n,P) \longrightarrow 0$, then $\int x dP_n \longrightarrow \int x dP$, $\int x^2 dP_n \longrightarrow \int x^2 dP$ and $\{P_n: n \geq 1\}$ converges weakly to P. The Lindeberg condition for an array of random variables ensures the weak convergence in L_2 to $\mathcal{N}(0,1)$. It is not hard to check these statements (Exercise). For example, the convergence results in Theorems 3.4.1, 3.1.1 and 1.3.1 are in L_2.

We shall now prove the asymptotic normality for simple linear rank statistics under alternatives.

Theorem 3.4.3: For every $N \in \mathbb{N}$ let $X_{1N},\ldots,X_{NN}$ be independent random variables with continuous d.f. $F_{1N},\ldots,F_{NN}$, and let $C_{1N},\ldots,C_{NN}$ be regression constants satisfying

$$\sup_{i,N} |C_{iN}| \leq 1.$$

Then for every absolutely continuous proper score function $h \in \mathfrak{G}$ we have

$$(3.4.14) \qquad \lim_{N\to\infty} D_2(\mathcal{L}(T_N(h)), \mathcal{N}(0,\sigma_N^2(h))) = 0$$

where $\sigma_N^2(h)$ is given by (3.4.10).

Remark: The convergence in the preceding theorem is uniform for all continuous families $\{F_{iN}: 1 \leq i \leq N, N \in \mathbb{N}\}$ and for all $\{C_{iN}: 1 \leq i \leq N, N \in \mathbb{N}\}$ with $\sup |C_{iN}| \leq 1$. It is also uniform over bounded classes of absolutely continuous proper score functions. This follows from the proof below. (Exercise)

Proof: Adding a constant c to each of the regression constants changes $T_N(h)$ by at most

$$N^{1/2} c \left| N^{-1} \sum_{i=1}^{N} h(i/N+1) - \int h(t)\, dt \right|.$$

Decomposing $h = h_1 - h_2$ according to Lemma 3.3.2 and using Lemma 3.3.4 (3.3.12) it is easy to see that

$$N^{1/2} \left(N^{-1} \sum_{1 \leq i \leq N} h(i/N+1) - \int h(t)\, dt\right) \to 0 .$$

Hence we may assume that $C_{iN} \geq 0 \quad (1 \leq i \leq N,\ N \in \mathbb{N})$.

We shall prove the theorem for $h \in C_{2b}$ first using Taylor expansion. The general case follows then from Theorem 3.4.2 and Lemma 3.3.5 .

Let $h \in C_{2b}$. Using Taylor's theorem at $H_N(t)$ we obtain (cf. (3.4.8),(3.4.9))

$$\begin{aligned}
T_N(h) &= N^{1/2} \int_{-\infty}^{\infty} h(H_N(t))\, d(\hat{F}_N - F_N)(t) \\
&\quad + N^{1/2} \int_{-\infty}^{\infty} h'(H_N(t)) \left(\tfrac{N}{N+1} \hat{H}_N(t) - H_N(t)\right) d\hat{F}_N(t) \\
&\quad + \tfrac{1}{2} N^{1/2} \int_{-\infty}^{\infty} h''(\vartheta_N(\hat{H}_N(t)) \left(\tfrac{N}{N+1} \hat{H}_N(t) - H_N(t)\right)^2 d\hat{F}_N(t) \qquad (3.4.15) \\
&= W_N(h) + A - B + C ,
\end{aligned}$$

where

$$A := N^{1/2} \int_{-\infty}^{\infty} h'(H_N(t))\left(\tfrac{N}{N+1} \hat{H}_N(t) - H_N(t)\right) d(\hat{F}_N - F_N)(t)$$

$$B := \frac{\sqrt{N}}{N+1} \int_{-\infty}^{\infty} h'(H_N(t))\hat{H}_N(t)\, dF_N(t)$$

$$C := \frac{1}{2} \sqrt{N} \int_{-\infty}^{\infty} h''(\vartheta_N(\hat{H}_N(t)))\left(\tfrac{N}{N+1} \hat{H}_N(t) - H_N(t)\right)^2 d\hat{F}_N(t).$$

Lemma 3.3.6, relations (3.3.17) and (3.3.18), yield that for some universal constant c

$$EA^2 \leq cN^{-1} \int h'^2(t)\, dt$$

$$EB^2 \leq cN^{-1} \int h'^2(t)\, dt$$

and $EC^2 \leq 4^{-1} cN\, E\left(\int_{-\infty}^{\infty} \left(\tfrac{N}{N+1} \hat{H}_N(t) - H_N(t)\right)^2 d\hat{H}_N(t)\right)^2$

$$\leq \quad \frac{cN}{4} \sum_{l=0}^{2} N^{-4+l} \int (t(1-t))^{l}\, dt \leq \frac{3c}{4N} \quad .$$

Therefore $E(T_N(h)-W_N(h))^2 \longrightarrow 0$ as $N\to\infty$. Since $W_N(h)$ is a sum of independent, bounded random variables, the Lindeberg condition is satisfied and therefore

$$\lim_{N\to\infty} D_2 \left(\mathcal{L}(W_N(h)), \mathcal{N}(0,\sigma_N^{\,2}(h))\right) = 0.$$

Consequently

$$\lim_{N\to\infty} D_2 \left(\mathcal{L}(T_N(h)), \mathcal{N}(0,\sigma_N^{\,2}(h))\right) = 0,$$

and (3.4.14) holds for $h\in C_{2b}$.

Let $h \in \mathfrak{G}$ be arbitrary. Using Theorem 3.4.2, (3.4.7), and the triangle inequality for D_2 we have for $g\in C_{2b}$

$$D_2(\mathcal{L}(T_N(h)),\mathcal{N}(0,\sigma_N^{\,2}(h))) \leq D_2(\mathcal{L}(T_N(h)),\mathcal{L}(T_N(g))) +$$

$$+D_2(\mathcal{L}(T_N(g)),\mathcal{N}(0,\sigma_N^{\,2}(g))) + D_2(\mathcal{N}(0,\sigma_N^{\,2}(g)),\mathcal{N}(0,\sigma_N^{\,2}(h)))$$

$$\leq K \,\| h-g \|^{2}+(\sigma_N(g)-\sigma_N(h))^2+D_2(\mathcal{L}(T_N(g)),\mathcal{N}(0,\sigma_N^{\,2}(g))).$$

Here we used the following fact: If X is $\mathcal{N}(0,\sigma^2)$ distributed, then $\tau\sigma^{-1}X$ is $\mathcal{N}(0,\tau^2)$ distributed and $E(\tau\sigma^{-1}X-X)^2=\sigma^2(\tau\sigma^{-1}-1)^2 =(\tau-\sigma)^2$. Therefore $D_2(\mathcal{N}(0,\sigma^2),\mathcal{N}(0,\tau^2)) \leq (\tau-\sigma)^2$.

Since by Lemma 3.3.5 h can be approximated by C_{2b}-functions, (3.4.14) follows if we can show that $h\to \sigma_N^{\,2}(h)$ is a uniformly (in N) continuous family of functions.

In fact we shall show that $\sigma_N\colon \mathbb{H}\times\mathbb{H}\to\mathbb{R}$, defined by

$$(3.4.16) \quad \sigma_N(f,g)= \frac{1}{N} \sum_{i=1}^{N} \iint_{s<t} F_{iN}(s)(1-F_{iN}(t))\left(\frac{dF_N}{dH_N}(t)-C_{iN}\right)$$

$$\left(\frac{dF_N}{dH_N}(s) - C_{iN}\right) \left[dg(H_N(t))df(H_N(s)) + df(H_N(t))dg(H_N(s))\right]$$

are uniformly bounded operators.

Let f and g be increasing. Since $\frac{dF_N}{dH_N}$ is bounded and since for $s<t$ by Cauchy- Schwarz' inequality

$$N^{-1} \sum_{i=1}^{N} F_{iN}(s)(1-F_{iN}(t)) \leq N^{-1} \sum_{i=1}^{N} [F_{iN}(s)(1-F_{iN}(s)) F_{iN}(t)(1-F_{iN}(t))]^{1/2}$$

$$\leq (H_N(s)(1-H_N(s)))^{1/2}(H_N(t)(1-H_N(t)))^{1/2} \quad (\text{cf. } (3.3.19)),$$

it follows for some constant c that

$$\sigma_N(f,g) \leq c \iint_{s<t} (H_N(s)(1-H_N(s)))^{1/2}(H_N(t)(1-H_N(t)))^{1/2} [df(H_N(t))dg(H_N(s))+df(H_N(s))dg(H_N(t))]$$

$$\leq c \; \|f\| \, \|g\|.$$

If f and g are arbitrary, then by Lemma 3.3.5 and linearity of σ_N, $|\sigma_N(f,g)| \leq c \; \|f\| \, \|g\|$.

This completes the proof of the theorem.

Corollary: If we replace in the previous theorem $T_N(h)$ by

$$S_N = N^{1/2}\left(\int h\left(\tfrac{N}{N+1}\hat{H}_N(t)\right)d\hat{F}_N(t) - E\left(\int h\left(\tfrac{N}{N+1}\hat{H}_N(t)\right) d\hat{F}_N(t)\right)\right)$$

then

$$\lim_{N\to\infty} D_2(\mathcal{L}(S_N), \mathcal{N}(0,\sigma_N^2(h))) = 0.$$

Proof: We have to show that

$$\lim N^{1/2}\left(E\left(\int h\left(\tfrac{N}{N+1}\hat{H}_N(t)\right) d\hat{F}_N(t)\right) - \int h(H(t)) \, dF_N(t)\right) = 0.$$

If h belongs to $C_{2,b}$, then this follows using Taylor expansion and Lemma 3.3.6, (3.3.17). Then approximate a general $h \in \mathcal{G}$ by $g \in C_{2,b}$ and use Theorem 3.4.2, (3.4.7). We leave the details as an exercise.

Theorem 3.4.3 is directly applicable to the van der Waerden two sample linear rank statistic, but not, for example, to the Fisher-Yates-Terry-Hoeffding normal score statistic, for which we finally give a result.

Theorem 3.4.4: Let h be an absolutely continuous proper score function satisfying $h''(t) \leq (t(1-t))^{-5/2+\delta}$ for some $\delta>0$. Define $a_N(i) = Eh(U_{(i)})$ where $U_{(1)}<U_{(2)}<\ldots<U_{(N)}$ denotes the order statistic of N i.i.d. uniformly distributed random variables. Let C_{iN} denote regression constants satisfying $\sup_{i,N}|C_{iN}|\leq 1$ and let $\{X_{iN}:1\leq i\leq N\}$ denote independent random variables with continuous d.f. F_{iN} $(N\geq 1)$.
Define $T_N=N^{-1} \sum_{1\leq i\leq N} C_{iN}a_n(R_{iN})$. Then $\sqrt{N}(T_N- \int h(H_N(t))\, dF_N(t))$

converges weakly in L_2 to $N(0,\sigma_N^{\,2}(h))$, where $\sigma_N^{\,2}(h)$ is defined in (3.4.10).

Proof: Obviously, by Theorem 3.4.3 it suffices to show that

$$\lim_{N\to\infty} E(N^{1/2}(T_N- \int_{-\infty}^{\infty} h(\tfrac{N}{N+1}\, \hat{H}_N(t))\, d\hat{F}_N(t)))^2 = 0.$$

Since

$$|T_N- \int_{-\infty}^{\infty} h(\tfrac{N}{N+1}\, \hat{H}_N(t))d\hat{F}_N(t)| \leq N^{-1} \sum_{i=1}^{N} |a_N(i)-h(\tfrac{i}{N+1})|,$$

we have to prove that

$$\lim_{N\to\infty} N^{-1/2} \sum_{i=1}^{N} |Eh(U_{(i)})-h(\tfrac{i}{N+1})| = 0. \tag{3.4.17}$$

The d.f. of $U_{(i)}$, is given by

$$\Phi_i(s):=P(U_{(i)}\leq s)= \sum_{k\geq i} \binom{N}{k} s^k(1-s)^{N-k}=i\binom{N}{i} \int_0^s t^{i-1}(1-t)^{N-i}\, dt$$

and therefore Φ_i has density $\varphi_i(s)=i\binom{N}{i}s^{i-1}(1-s)^{N-i}$. But this is a Beta-distribution with parameters $\alpha=i$ and $\beta=N-i+1$. It follows that $EU_{(i)}=\frac{\alpha}{\alpha+\beta} = \frac{i}{N+1}$ and $\operatorname{Var} U_{(i)} = \frac{\alpha\beta}{(\alpha+\beta)^2(\alpha+\beta+1)} =$

$= ((N+2)(N+1))^{-1}i(1-i(N+1)^{-1})$. Observe that

$$A_1 := N^{-1/2} \sum_{i=1}^{N} Eh(U_{(i)})1_{\{U_{(i)} \leq \frac{1}{N+1}\}} =$$

$$= N^{1/2} \int_0^{\frac{1}{N+1}} h(t) \sum_{i=0}^{N-1} \binom{N-1}{i}t^i(1-t)^{N-1-i}\, dt =$$

$$= N^{1/2} \int_0^{1/(N+1)} h(t)\, dt \longrightarrow 0 \qquad \text{as } N \longrightarrow \infty$$

and, similarly, as $N \to \infty$,

$$A_2 := N^{-1/2} \sum_{i=1}^{N} Eh(U_{(i)})\, 1_{\{U_{(i)} \geq \frac{N}{N+1}\}} \to 0,$$

$$A_3 := N^{-1/2} \sum_{i=1}^{N} h'(\tfrac{i}{N+1}) \int_0^{\frac{1}{N+1}} |t - \tfrac{i}{N+1}|\, d\Phi_i(t) \to 0,$$

$$A_4 := N^{-1/2} \sum_{i=1}^{N} h'(\tfrac{i}{N+1}) \int_{\frac{N}{N+1}}^{1} |t - \tfrac{i}{N+1}|\, d\Phi_i(t) \to 0$$

and

$$A_5 := N^{-1/2} \sum_{i=1}^{N} h(\tfrac{i}{N+1})\, (1 - \Phi_i(\tfrac{N}{N+1}) + \Phi_i(\tfrac{1}{N+1})) \to 0.$$

Using Taylor's theorem, (3.4.17) is bounded by

$$(3.4.18) \qquad \sum_{j=1}^{5} A_j + N^{-1/2} \sum_{i=1}^{N} \Big\{ |h'(\tfrac{i}{N+1}) E(U_{(i)} - \tfrac{i}{N+1})| + \int_{\frac{1}{N+1}}^{\frac{N}{N+1}} |h''(\vartheta_{i,t})|\, (t - \tfrac{i}{N+1})^2\, d\Phi_i(t) \Big\},$$

where $\vartheta_{i,t}$ belongs to the interval between t and $\frac{i}{N+1}$. Since

$|h''(\vartheta_{i,t})| \leq (t(1-t))^{-5/2+\delta} + (\frac{i}{N+1}(1-\frac{i}{N+1}))^{-5/2+\delta}$ by assumption,

we have

$$N^{-1/2} \sum_{i=1}^{N} \int_{\frac{1}{N+1}}^{\frac{N}{N+1}} (\tfrac{i}{N+1}(1 - \tfrac{i}{N+1}))^{-5/2+\delta} (t - \tfrac{i}{N+1})^2\, d\Phi_i(t) \leq$$

$$\leq N^{-1/2} (N+2)^{-1} \sum_{i=1}^{N} (\tfrac{i}{N+1}(1 - \tfrac{i}{N+1}))^{-3/2+\delta} \leq$$

$$\leq \text{const. } N^{-\delta} \longrightarrow 0$$

and

$$N^{-1/2} \sum_{i=1}^{N} \int_{\frac{1}{N+1}}^{\frac{N}{N+1}} (t(1-t))^{-5/2+\delta} (t - \tfrac{i}{N+1})^2\, d\Phi_i(t) =$$

$$\leq N^{-\delta/2} \int_{\frac{1}{N+1}}^{\frac{N}{N+1}} (t(1-t))^{-2+\delta/2}\, N \sum_{i=0}^{N-1} (\tfrac{i+1}{N+1} - t)^2 \binom{N-1}{i} t^i (1-t)^{N-1-i}\, dt =$$

$$= N^{-\delta/2} \int_{\frac{1}{N+1}}^{\frac{N}{N+1}} (t(1-t))^{-2+\delta/2} \frac{N}{(N+1)^2} [(N-1)t(1-t)+(1-2t)^2] \, dt \to 0.$$

Combining all estimates the upper bound (3.4.18) must converge to zero, proving (3.4.17) and the theorem.

Exercise: Show that the v.d. Waerden score function Φ^{-1} satisfies the assumption on the score function in the last theorem. This proves e.g. the asymptotic normality of the Fisher-Yates-Terry Hoeffding normal score statistic in Example 3.2.1 .

5. Signed rank statistics and R-estimators

We extend the method introduced in the last two sections to signed rank statistics and derive the asymtotic normality for R-estimators.

Definition 3.5.1: A statistic $T:\mathbb{R}^n \longrightarrow \mathbb{R}$ is called a *signed rank statistic* if T can be written as

$$(3.5.1) \qquad T(x) = \sum_{i=1}^{n} C_i \, a(r_i^+) \operatorname{sign} x_i \qquad (x=(x_1,\ldots,x_n)\in\mathbb{R}^n)$$

where C_i denote *regression constants* , a(i) *scores* and r_i^+ the rank of $|x_i|$ among $|x_1|,\ldots,|x_n|$.

Signed rank statistics are mostly used to test for symmetry. Under the hypothesis (i.e. F is continuous and symmetric about $\vartheta_o=0$; $X_1,\ldots,X_n$ are i.i.d. F-distributed) the events $\{R_i^+=r_i, \operatorname{sign} X_i=\varepsilon_i \ 1\le i\le n\}$ have probability $(n!)^{-1}2^{-n}$ and hence

$$E\, a(R_i^+) \operatorname{sign} X_i = 0$$

and

$$E\, a(R_i^+)\, a(R_j^+) \operatorname{sign} X_i \operatorname{sign} X_j =$$

$$= \begin{cases} \displaystyle\sum_{u,u'=1}^{n} \sum_{\varepsilon_i,\varepsilon_j\in\{\pm 1\}} \varepsilon_i\varepsilon_j \, a(u)a(u') \frac{1}{4n(n-1)} = 0 & (i\neq j) \\ \displaystyle\sum_{u=1}^{n} \sum_{\varepsilon\in\{\pm 1\}} \varepsilon^2 \, a(u)^2 \frac{1}{2n} = n^{-1} \sum_{u=1}^{n} a(u)^2 & (i=j). \end{cases}$$

Thus we have proved

Lemma 3.5.1: Under the above hypothesis we have

$$ET = 0 \tag{3.5.2}$$

$$\text{Var } T = \sum_{i=1}^{n} c_i^2 a(i)^2 \tag{3.5.3}$$

and

$$E \sum_{i=1}^{n} a(R_i^+) = n\,\bar{a}\,. \tag{3.5.4}$$

Very often a signed rank statistic $T = \sum a(r_i^+)$ sign x_i appears transformed into

$$T_1 = 1/2(T + n\bar{a}) = 1/2 \sum_{i=1}^{n} a(r_i^+)(\text{sign } x_i + 1)$$

$$= \sum_{i:x_i>0} a(r_i^+)\,. \tag{3.5.5}$$

In this case $ET_1 = 1/2\, n\bar{a}$ and $\text{Var } T_1 = 1/4 \sum_{i=1}^{n} a(i)^2$.

Let us also note the following easy fact.

Lemma 3.5.2: Let F be a symmetric, continuous d.f. and let $X_1,\ldots,X_n$ be i.i.d. F-distributed. Then $F^{-1}[(1/2)(1+U_{(i)})]$ has the same distribution as $|X|_{(i)}$, where $|X|_{(1)} < |X|_{(2)} < \ldots < |X|_{(n)}$ denotes the order statistic of $(|X_j|)$ and where $U_{(i)}$ $(1 \le i \le n)$ is the order statistic of n i.i.d. uniformly distributed random variables.

Proof: Let $s > 0$

$$P(F^{-1}(1/2 + [1/2]U_{(i)}) \le s) = P(U_{(i)} \le 2F(s) - 1)$$

$$= \sum_{k \ge i} \binom{n}{k} (2F(s)-1)^k (2-2F(s))^{n-k}$$

and

$$P(|X|_{(i)} \le s) = \sum_{k \ge i} P(|\{j : -s \le X_j \le s\}| = k)$$

$$= \sum_{k \ge i} \binom{n}{k} (2F(s)-1)^k (2-2F(s))^{n-k},$$

since $P(-s \le X_1 \le s) = 2F(s) - 1$ by symmetry.

Example 3.5.1: The Fraser normal score test uses the regression constants $c_i = 1$ and the scores $a(i) = E\Phi^{-1}[(1/2)(1+U_{(i)})]$ $(1 \le i \le n)$, where Φ denotes the d.f. of $\mathcal{N}(0,1)$ and where $U_{(i)}$

denotes the order statistic of n i.i.d. uniformly distributed random variables. If the d.f. of X_1 is symmetric about 0 then $ET = 0$, $\text{Var } T = \sum a(i)^2$ and

$$\sum_{i=1}^{n} a(i) = \sum_{i=1}^{n} \int_0^1 \Phi^{-1}(1/2 + t/2) i \binom{n}{i} t^{i-1}(1-t)^{n-i}\, dt =$$

$$= 2n \int_0^{\infty} u\, d\Phi(u) = n\sqrt{\frac{2}{\pi}}.$$

For the v.d. Waerden normal score test take $a(i)=\Phi^{-1}(1/2+ +1/2 \frac{i}{n+1})$; expectation and variance are given by Lemma 3.5.1, again under symmetry assumptions.

The Wilcoxon signed rank statistic is defined by

$$T_1 = \sum_{i:X_i>0} R_i^+ = 1/2\left(\sum_{i=1}^{n} R_i^+ \text{ sign } X_i + \frac{n(n+1)}{2}\right)$$

(cf. (3.5.5)). Hence $ET_1 = \frac{n(n+1)}{4}$ and $\text{Var } T_1 = 1/24\, n(n+1)(2n+1)$ under symmetry. This statistic is commonly used.
Finally, the sign test is given by

$$T_1 = \sum_{i:X_i>0} 1 = 1/2\left(\sum_{i=1}^{n} \text{sign } X_i + n\right)$$

with scores $a(i) = 1$. Under symmetry, $ET_1 = \frac{n}{2}$ and $\text{Var } T_1 = \frac{n}{4}$.

Some of the above test statistics are locally most powerful under certain location alternatives according to the following theorem.

Theorem 3.5.1: Let $\{P_\vartheta : \vartheta \in I\}$ be a location problem, where I is some right open interval containing 0. Denote by $F_\vartheta(x)=F(x-\vartheta)$ the d.f. of P_ϑ and assume that $F=F_0$ has an absolutely continuous symmetric density f for which f' is Lebesgue integrable.

For $N \in \mathbb{N}$ let $T=T_N= \sum_{i=1}^{N} a(r_i^+) \text{ sign } X_i$ denote the signed rank statistic with scores

(3.5.6) $\quad a(i)=E\left(-\frac{f'}{f} \circ F^{-1}(1/2 + 1/2\, U_{(i)}\right) \quad (1 \leq i \leq N).$

Then for any $0<\alpha<1$ there exists a $k = k(\alpha)$ such that the test with critical region $\{T>k\}$ is a locally most powerful rank test for $H=\{Q_0:=P_0^N\}$ against $K=\{Q_\vartheta:=P_\vartheta^N:\vartheta>0\}$.

Proof: We proceed similarly to the proof of Theorem 3.2.1. Fix $\vartheta>0$. From the Neyman-Pearson lemma it follows that the most powerful rank test φ for $\vartheta_0=0$ against ϑ has the form

$$\{\varphi=1\}=\{(r,\varepsilon)=(r_1^+,\dots,r_N^+,\varepsilon_1,\dots,\varepsilon_N) : Q_\vartheta(r,\varepsilon) > \frac{K}{N!2^N}\}$$

where K is chosen by

$$|\{(r,\varepsilon):Q_\vartheta(r,\varepsilon)>K(N!)^{-1}2^{-N}\}| \leq \alpha N!2^N < |\{(r,\varepsilon):Q_\vartheta(r,\varepsilon)\geq K(N!)^{-1}2^{-N}\}|.$$

(Here r denotes a rank vector and ε a sign vector.) Therefore we have to show that for different ϑ (but small enough) the same rank-sign-vectors form the critical region and that these vectors are determined by $\sum a(r_i^+)\varepsilon_i>k$ for some $k=k(\alpha)$. This follows from the following statement: There exists an $a>0$ such that the inequality

$$\sum a(r_i^+)\varepsilon_i > \sum a(r_i'^+)\varepsilon_i'$$

implies for any $0<\vartheta<a$

$$Q_\vartheta(\{x:r(x)=r,\varepsilon(x)=\varepsilon\})>Q_\vartheta(\{x:r(x)=r',\varepsilon(x)=\varepsilon'\}).$$

In order to prove this statement, let $A=\{x:r(x)=r,\varepsilon(x)=\varepsilon\}$ for fixed $r=(r_1^+,\dots,r_n^+)$ and $\varepsilon=(\varepsilon_1,\dots,\varepsilon_N)$. Since $f(x-\vartheta)$ is the density of F_ϑ,

$$\begin{aligned} Q_\vartheta(A)&=Q_0(A) + Q_\vartheta(A) - Q_0(A)\\ &=(N!)^{-1}2^{-N}+\vartheta \sum_{k=1}^{N} \int\dots\int_A \frac{f(x_k-\vartheta)-f(x_k)}{\vartheta} \prod_{i=1}^{k-1} f(x_i-\vartheta) \prod_{i=k+1}^{N} f(x_i) \\ &\qquad dx_1\dots dx_N. \end{aligned}$$

For fixed $k=1,\dots,N$ define

$$g_\vartheta(x_1,\dots,x_N)=\vartheta^{-1}(f(x_k-\vartheta)-f(x_k)) \prod_{i=1}^{k-1} f(x_i-\vartheta) \prod_{i=k+1}^{N} f(x_i)$$

and

$$g(x_1,\dots,x_N) = -f'(x_k) \prod_{i \neq k} f(x_i).$$

We claim that

$$(3.5.7) \qquad \lim_{\vartheta \to 0} \int_A g_\vartheta \, d\lambda^N = \int_A g d\lambda^N.$$

If this holds for each $k=1,\dots,N$, the proof of the theorem is complete, because of the following argument:
From the invariance of Q_o under all mappings ψ of the form $(x_1,\dots,x_N) \longrightarrow (\varepsilon_1' \, x_{\tau(1)},\dots,\varepsilon_N' \, x_{\tau(N)})$, where $\tau \in \gamma_N$ and $\varepsilon_i' \in \{\pm 1\}$, one obtains

$$\int_A g d\lambda^N = -\int_A \frac{f'(x_k)}{f(x_k)} \, dQ_o(x_1,\dots,x_N) =$$

$$= -(N!)^{-1} 2^{-N} \sum_\psi \int_{\psi^{-1}A} \frac{f'(\varepsilon_k' \, x_{\tau(k)})}{f(\varepsilon_k' \, x_{\tau(k)})} \, dQ_o(x_1,\dots,x_N).$$

If $x \in \psi^{-1}A$ then $|x_{\tau(k)}|$ has rank r_k^+ among $|x_1|,\dots,|x_N|$, where r_k^+ was fixed above.

Denote by $|x|_{(r_k^+)}$ this absolute value. Also sign $\varepsilon_k' \, x_{\tau(k)} = \varepsilon_k$ and hence $\varepsilon_k' \, x_{\tau(k)} = \varepsilon_k \, |x|_{(r_k^+)}$. Using $f'(x) = -f'(-x)$ and $f(x) = f(-x)$, it follows that

$$\int_A g d\lambda^N = -(N!)^{-1} 2^{-N} \varepsilon_k \sum_\psi \int_{\psi^{-1}A} \frac{f'(|x|_{(r_k^+)})}{f(|x|_{(r_k^+)})} \, dQ_o(x_1,\dots,x_N)$$

$$= -(N!)^{-1} 2^{-N} \varepsilon_k \, E \, \frac{f'(|X|_{(r_k^+)})}{f(|X|_{(r_k^+)})} = (N!)^{-1} 2^{-N} \varepsilon_k \, a(r_k^+)$$

by Lemma 3.5.2.

It remains to prove (3.5.7). By Fubini's theorem

$$\int |g_\vartheta| \, d\lambda^N = \int_{-\infty}^{\infty} \vartheta^{-1} \left| \int_0^\vartheta f'(x-\delta) \, d\delta \right| dx \leq \int |f'| \, d\lambda = \int |g| \, d\lambda^N.$$

Since $g_\vartheta \to g$ λ^N-almost surely as $\vartheta \to 0$, Fatou's lemma implies that

$$\liminf_{\vartheta \longrightarrow 0} \int_C g_\vartheta^\pm \, d\lambda^N \geq \int_C g^\pm \, d\lambda^N$$

for the positive (negative) parts of g_ϑ and g and for each measurable set C. Finally, if for some sequence $\vartheta_j \to 0$ and some measurable set C

$$\lim_j \int_C g^+_{\vartheta_j} d\lambda^N > \int_C g^+ \, d\lambda^N$$

then

$$\limsup_j \int |g_{\vartheta_j}| \, d\lambda^N$$

$$\geq \lim_j \int_C g^+_{\vartheta_j} \, d\lambda^N + \liminf \int_{C^c} g^+_{\vartheta_j} d\lambda^N +$$

$$+ \liminf \int g^-_{\vartheta_j} \, d\lambda^N$$

$$> \int |g| \, d\lambda^N.$$

This contradiction proves (3.5.7) together with similar arguments for g^-_ϑ and g^-.

For the asymptotic normality of signed rank statistics we shall make use of sections 3 and 4. The proof works in two steps. We first reduce the problem to a two-sample case and then apply the previous method.

Lemma 3.5.3:

Let $X_1,\dots,X_n$ be random variables with continuous d.f. F. Denote by R^+_i the rank of $|X_i|$ among $|X_1|,\dots,|X_n|$ and by R_i the rank of X_i $(i \leq n)$, [resp. of $-X_{i-n}$] among all values of $X_1,\dots,X_n, -X_1,\dots,-X_n$. Then for $i \leq n$.

$$R^+_i = \begin{cases} R_i - n & \text{if } X_i > 0 \\ n+1-R_i & \text{if } X_i < 0. \end{cases} \tag{3.5.8}$$

In particular, if $T = \sum_{i=1}^n a(R^+_i) \operatorname{sign} X_i$ is a signed rank statistic, then T has a representation

$$T = \sum_{i=1}^{2n} C_i \, b(R_i) \tag{3.5.9}$$

where $C_i = 1$ if $i \leq n$ and $C_i = 0$ if $i > n$ and where

$$b(i) = \begin{cases} a(i-n) & \text{if } i > n \\ -a(n+1-i) & \text{if } i \leq n. \end{cases} \tag{3.5.10}$$

Proof:

If $X_i > 0$, then

$$R_i^+ = \sum_{j=1}^{n} 1_{\{-X_i<X_j\leq X_i\}} = \sum_{j=1}^{n} (1_{\{X_j\leq X_i\}} -1+1_{\{-X_j<X_i\}}) = R_i-n.$$

If $X_i < 0$, then

$$R_i^+ = 1 + \sum_{j=1}^{n} 1_{\{X_i<X_j\leq -X_i\}} =$$

$$= 1 + \sum_{j=1}^{n} (1-1_{\{X_j\leq X_i\}} - 1_{\{-X_j<X_i\}}) = n+1-R_i.$$

Example 3.5.2:

Let the scores $a(i)$ be given by $a(i) = h(1/2+1/2\,\frac{i}{n+1})$. Then

$$b(i) = \begin{cases} a(i-n) = h(\frac{i}{N+2} + \frac{1}{N+2}) = g_N(\frac{i}{N+1}) & \text{if } i > n \\ -a(n+1-i)=-h(-\frac{i}{N+2}+1) = g_N(\frac{i}{N+1}) & \text{if } i \leq n \end{cases}$$

where

$$g_N(t) = \begin{cases} h(\frac{N+1}{N+2}\,t + \frac{1}{N+2}) & (t \geq 1/2) \\ -h(1-\frac{N+1}{N+2}\,t) & (t < 1/2 \end{cases} \quad \text{and where } N = 2n.$$

Setting

$$g(t) = \begin{cases} h(t) & \text{if } t \geq 1/2 \\ -h(1-t) & \text{if } t < 1/2, \end{cases}$$ observe that for absolutely continuous $h \in \mathfrak{C}$

$$g'(t) = \begin{cases} h'(t) & t \geq 1/2 \\ h'(1-t) & t < 1/2 \end{cases}$$

and therefore $\int_0^1 \sqrt{t(1-t)}\ |g'(t)|\ dt =$

$$= 2 \int_{1/2}^1 \sqrt{s(1-s)}\ |h'(s)|\ ds < \infty.$$

It follows that $g \in \mathfrak{C}$.

Assume that h is nondecreasing. Then $g_N(t) \geq g(t)$ for $t \geq 1/2$ and $g_N(t) \leq g(t)$ for $t < 1/2$. Hence

$$\sum_{i=1}^{n} \left| g_N(\frac{R_i}{N+1}) - g(\frac{R_i}{N+1}) \right| = \sum_{i:R_i>n} g_N(\frac{R_i}{N+1}) - g(\frac{R_i}{N+1}) +$$

$$+ \sum_{i:R_i \leq n} g(\frac{R_i}{N+1}) - g_N(\frac{R_i}{N+1}) \leq$$

$$\leq \sum_{i=1}^{n} g(\frac{i}{N+1}) - g_N(\frac{i}{N+1}) + \sum_{i=n+1}^{N} g_N(\frac{i}{N+1}) - g(\frac{i}{N+1}) =$$

$$= \sum_{i=1}^{n} h(1-\frac{i}{N+2}) - h(1-\frac{i}{N+1}) + \sum_{i=n+1}^{N} h(\frac{i+1}{N+2}) - h(\frac{i}{N+1}) \leq$$

$$\leq 2(h(1-\frac{1}{N+2}) - h(1-\frac{n}{N+1})).$$

If $h \in \mathcal{H}$, then by Lemma 3.3.4, (3.3.12),

$$\text{(3.5.11)} \quad \lim_{N\to\infty} N^{-1/2} \sum_{i=1}^{n} \left| g_N(\frac{R_i}{N+1}) - g(\frac{R_i}{N+1}) \right| = 0 .$$

This example shows how to reduce the problem of asymptotic normality of signed rank statistics to the case of a two-sample linear rank statistic. However, the two-samples are no longer independent. If $X_1,\dots,X_n$ are i.i.d. with continuous d.f. F define the second sample by $-X_1,\dots,-X_n$. We shall use the same notation for $\hat{F}_N$, $\hat{H}_N$, F_N, H_N etc., where $N = 2n$, as in sections 3 and 4 ((3.3.2), (3.3.3), (3.3.10), (3.4.8)), especially

$$\text{(3.5.12)} \quad T_N(h) = N^{1/2}\left(\int_{-\infty}^{\infty} h(\tfrac{N}{N+1}\hat{H}_N(t))\,d\hat{F}_N(t) - \int_{-\infty}^{\infty} h(H_N(t))\,dF_N(t)\right)$$

denotes the signed rank statistic which is of interest to us. First observe that $2F_N = F$ is just the distribution of X_1 and that $H_N(t) = N^{-1}\sum_{i=1}^{n}(P(X_i \leq t) + P(X_i \geq -t)) = 1/2(1+F(t)-F(-t)$, if the regression constants are as in Example 3.5.2. However, the C_{iN} need not necessarily be given by 0 and 1. We proceed as in Theorems 3.4.2 and 3.4.3.

<u>Theorem 3.5.2:</u>

There exists a constant $K > 0$ such that for every i.i.d. sequence $X_1,\dots,X_n$ with continuous d.f. F, every array of regression constants C_{iN} $(1 \leq i \leq N)$ with $\max_i |C_{iN}| \leq 1$ and every $h \in \mathcal{H}$

$$(3.5.13) \quad E\, T_N(h)^2 \leq K \|h\|^2 .$$

Proof:

The proof of Theorem 3.4.2 remains unchanged in the present situation with two exceptions. The estimate for $E|\hat{H}(H^{-1}(\frac{N+1}{N} z)-) - \frac{N+1}{N} z|^2$ in the proof of (3.4.4) is replaced by

$$E(\hat{H}_N(H_N^{-1}(\frac{N+1}{N} z)-) - \frac{N+1}{N} z)^2$$

$$= E(N^{-1} \sum_{i=1}^{n} (1_{\{X_i < H_N^{-1}(\frac{N+1}{N}z)\}} + 1_{\{-X_i < H_N^{-1}(\frac{N+1}{N}z)\}} - \frac{N+1}{n} z))^2$$

$$= n\, N^{-2} \frac{N+1}{N} z(1-\frac{N+1}{N} z) \leq N^{-1} z(1-z)$$

since

$$E\left[1_{\{X_i < H_N^{-1}(\frac{N+1}{N}z)\}} + 1_{\{-X_i < H_N^{-1}(\frac{N+1}{N} z\}}\right] = F(H_N^{-1}(\frac{N+1}{N} z))$$

$$+ 1-F(-H_N^{-1}(\frac{N+1}{N} z)) = 2H_N(H_N^{-1}(\frac{N+1}{N} z)) = \frac{N+1}{n} z.$$

In the proof of the analogue to (3.4.5) one can argue similarly. (Exercise: Check the proof of (3.5.13) in detail.)

Remark:

As in (3.4.9), define

$$(3.5.14) \quad W_N = W_N(h) = \sqrt{N}\{ \int_{-\infty}^{\infty} h(H_N(t))\, d(\hat{F}_N - F_N)(t) + $$

$$+ \int_{-\infty}^{\infty} (\hat{H}_N - H_N)(t) \frac{dF_N}{dH_N}(t)\, dh(H_N(t))\}$$

and denote by $\sigma_N^2 = \sigma_N^2(h)$ its variance. We want to give an expression for σ_N^2 and start just before the formula (3.4.10), setting $\varphi(t) = \frac{dF_N}{dH_N}(t)$:

$$\sigma_N^2 = N^{-1} \int_{-\infty}^{\infty} \int_{-\infty}^{\infty} E(\sum_{i=1}^{n} (\varphi(t) - C_{iN})(1_{\{X_i \leq t\}} - F(t)) +$$

$$+ (\varphi(t) - C_{n+i,N})(1_{\{X_i \geq -t\}} - (1-F(-t))))$$

$$(\sum_{i=1}^{n}(\varphi(s)-C_{iN})(1_{\{X_i\le s\}} - F(s)) +$$

$$+ (\varphi(s) - C_{n+i,N})(1_{\{X_i\ge -s\}} -$$

$$-(1-F(-s)))) \; d\,h(H_N(t)dh(H_N(s))$$

$$= N^{-1}\iint\sum_{i=1}^{n}\Big[E(\varphi(t)-C_{iN})(\varphi(s)-C_{iN})(1_{\{X_i\le t\}}-F(t))(1_{\{X_i\le s\}}-F(s))$$

$$+ E(\varphi(t)-C_{n+i,N})(\varphi(s)-C_{n+i,N})(1_{\{X_i\ge -t\}} -$$

$$- (1-F(-t)))(1_{\{X_i\ge -s\}} - (1-F(-s)))$$

$$+ E(\varphi(t)-C_{iN})(\varphi(s)-C_{n+i,N})(1_{\{X_i\le t\}}-F(t))(1_{\{X_i\ge -s\}}-$$

$$- (1-F(-s)))$$

$$+ E(\varphi(t)-C_{n+i,N})(\varphi(s)-C_{iN})(1_{\{X_i\ge -t\}}-$$

$$- (1-F(-t)))(1_{\{X_i\le s\}}-F(s))\Big]dh(H_N(t))dh(H_N(s))$$

$$= 2N^{-1}\iint_{s<t}\sum_{i=1}^{n}(\varphi(t)-C_{iN})(\varphi(s)-C_{iN})F(s)(1-F(t))dh(H_N(t))dh(H_N(s))$$

$$+ \frac{2}{N}\iint_{s<t}\sum_{i=1}^{n}(\varphi(t)-C_{n+i,N})(\varphi(s)-C_{n+i,N})\,F(-t)(1-F(-s))\,dh(H_N(t))\,dh(H_N(s))$$

$$(3.5.15)\qquad - \frac{2}{N}\iint_{t<-s}\sum_{i=1}^{n}(\varphi(t)-C_{iN})(\varphi(s)-C_{n+i,N})\,F(t)(1-F(-s))\,dh(H_N(t))\,dh(H_N(s))$$

$$-2N^{-1}\iint_{t>-s}\sum_{i=1}^{n}(\varphi(t)-C_{iN})(\varphi(s)-C_{n+i,N})F(-s)(1-F(t))dh(H_N(t))\,dh(H_N(s)).$$

Note that $H_N(t) = 1/2\,(F(t)+1-F(-t))$ does not depend on N.

<u>Theorem 3.5.3:</u>
Let $\{X_i : i\ge 1\}$ be i.i.d. random variables with continuous d.f. F and let $\{C_{iN} : 1\le i\le N,\ N=2n, n\in\mathbb{N}\}$ be regression constants satisfying $\sup_{i,N}|C_{iN}| \le 1$. Then for any absolutely continuous proper score function h

$$(3.5.16) \quad \lim_{N\to\infty} D_2(\mathcal{L}(T_N(h)), \mathcal{N}(0,\sigma_N^2(h))) = 0$$

where $\sigma_N^2(h)$ is given by (3.5.15).

Moreover, (3.5.16) holds uniformly over continuous d.f. F.

Proof: Let us first show that $h \longrightarrow \sigma_N^2(h)$ is a uniformly continuous family of functions (cf. (3.4.16)). Since $|F_N(t)| \leq H_N(t)$ the functions φ in (3.5.15) (which depend on N) are uniformly bounded. Therefore $\sigma_N^2(h) \leq c\,\|h\|^2$ follows from Lemma 3.3.4 (3.3.13) and

$$H_N(s)(1-H_N(t)) = 1/4\ (1-F(-s)+F(s))(1-F(t)+F(-t)).$$

In fact

$$F(s)(1-F(t)) \leq (1-F(-s)+F(s))(1-F(t)+F(-t)) =$$
$$= 4H_N(s)(1-H_N(t)),$$
$$F(-t)(1-F(-s)) \leq 4H_N(s)(1-H_N(t)),$$
$$F(-s)(1-F(t)) \leq (F(-s)(1-F(-s)))^{1/2}(F(t)(1-F(t)))^{1/2} \leq$$
$$\leq 4\ (H_N(s)(1-H_N(s)))^{1/2}(H_N(t)(1-H_N(t)))^{1/2}$$

if $t \geq -s$ and

$$F(t)(1-F(-s)) \leq (F(t)(1-F(t)))^{1/2}(F(-s)(1-F(-s)))^{1/2} \leq$$
$$\leq 4\ (H_N(s)(1-H_N(s)))^{1/2}(H_N(t)(1-H_N(t)))^{1/2}$$

if $t \leq -s$.

As in the proof of Theorem 3.4.3 we may assume that $c_{iN} \geq 0$ for all i and N. If $h \in C_{2b}$ then Taylor's theorem yields a representation as in (3.4.15) and it is sufficient to give the L_2-estimates for the expressions A, B and C in order to prove (3.5.16). Clearly, for some constant c $EB^2 \leq c\,N^{-1} \int h'^2(t)\,dt$ remains valid. Using Lemma 3.3.6 again and decomposing $\hat{H}_N$ and $\hat{F}_N$ into a sum of two empirical processes of i.i.d. random variables (for example

$$H_N(t) = (2n)^{-1} \sum_{i=1}^{n} 1_{\{X_i \leq t\}} + (2n)^{-1} \sum_{i=1}^{n} 1_{\{-X_i \leq t\}}$$

it is easy to see that $EA^2 \leq cN^{-1}$ and $EC^2 \leq cN^{-1}$. (Exercise: Give the details of this computation. In Lemma 3.3.6 $(N/N+1)\hat{H}(t)-H(t)$ may be replaced by $(N/N+1)\hat{H}(-t)-H(-t)$.)

The proof of (3.5.16) for arbitrary absolutely continuous $h \in \mathcal{G}$ is now completed as the proof of Theorem 3.4.3.

Example 3.5.3: (example 3.5.2 continued.) The preceding theorem shows that $D_2(\mathcal{L}(T_N(g)),\mathcal{N}(0,\sigma_N^2(g)))\longrightarrow 0$ if $h \in \mathcal{G}$ where $g(t) = h(t)$ if $t \geq 1/2$ and $g(t) = -h(1-t)$ if $t < 1/2$ and where h was assumed to be increasing. Also note that we are considering the special regression constants $C_{iN} = 1$ $(i \leq n)$ and $C_{iN} = 0$ $(i>n)$. From (3.5.11) it follows that

$$\lim_{N\to\infty} D_2(\mathcal{L}(T_N(g_N)),\mathcal{N}(0,\sigma_N^2(g))) = 0,$$

thus by Lemma 3.5.3

$$N^{1/2}(N^{-1}T - \int_{-\infty}^{\infty} g(H_N(t))\, dF_N(t))$$

is asymptotically equivalent to $\mathcal{N}(0,\sigma_N^2(g))$ in the D_2-metric.

First note that $2H_N(t) = F(t) + 1 - F(-t)$ is independent of N and that $F_N(t) = N^{-1} \sum_{1\leq i\leq N} C_{iN}F_{iN}(t) = 2^{-1}F(t)$ also does not depend on N.

Formula (3.5.15) for $\sigma_N^2(g)$ simplifies as well:

$$\begin{aligned}\sigma_N^2(g) = & \iint_{s<t} (\varphi(t)-1)(\varphi(s)-1)F(s)(1-F(t))\, dg(H_N(t))\, dg(H_N(s)) \\ & + \iint_{s<t} \varphi(t)\varphi(s)F(-t)(1-F(-s))\, dg(H_N(t))dg(H_N(s)) \\ & - \iint_{s<-t} (\varphi(t)-1)\varphi(s)F(t)(1-F(-s))\, dg(H_N(t))dg(H_N(s)) \\ & - \iint_{-s<t} (\varphi(t)-1)\varphi(s)F(-s)(1-F(t))\, dg(H_N(t))dg(H_N(s)))\end{aligned}$$

is independent of N, denoted by $\sigma^2(g)$. Thus a signed rank statistic $T = \sum a(R_i^+)$ sign X_i is asymptotically normal $\mathcal{N}(\mu,\sigma^2/N)$ with $\mu = (1/2) \int g((1/2)(F(t)+1-F(-t)))\, dF(t)$ and σ^2 as above, provided the scores a(i) satisfy our assumptions.

Under the hypothesis, i.e. assuming that F is symmetric or equivalently $F(t) = 1 - F(-t)$, we have $H_N(t) = F(t)$ and hence

$$\begin{aligned}\mu = (1/2) \int g(t)\, dt = & -1/2 \int_{t\leq 1/2} h(1-t)\, dt \\ & + 1/2 \int_{t\leq 1/2} h(t)\, dt = 0\end{aligned}$$

and $\varphi(t) = 1/2$. It follows that (use partial integration)

$$\sigma^2(g) = 1/2 \iint_{0 \le s < t \le 1} s(1-t)\, dg(s)dg(t) \; +$$

$$+1/4 \iint_{1-s>t} ts\, dg(s)dg(t) \; +$$

$$+1/4 \iint_{1-s<t} (1-s)(1-t)\, dg(t)dg(s) \; =$$

$$= 1/2 \int_0^1 (1-t)(tg(t)-\int_0^t g(s)\, ds)\, dg(t) \; +$$

$$+ 1/4 \int_0^1 (1-t)(\int_{1-t}^1 g(s)\, ds - tg(1-t))\, dg(t) \; +$$

$$+ 1/4 \int_0^1 t((1-t)g(1-t) - \int_0^{1-t} g(s)\, ds)\, dg(t) \; =$$

$$= 1/2 \int_0^1 t(1-t)g(t)\, dg(t) + 1/2 \int_0^1 (1-t)g^2(t)\, dt \; -$$

$$- 1/4 \int_0^1 g(t)g(1-t)\, dt.$$

Now $2 \int t(1-t)g(t)\, dg(t) = - \int g^2(t)\, dt + 2 \int tg^2(t)\, dt$ yields

$$\sigma^2(g) = 1/4 \int g^2(t)\, dt - 1/4 \int g(t)g(1-t)\, dt. \tag{3.5.17}$$

Expressing this variance by the original score function h (3.5.17) becomes

$$\sigma^2(g) = 1/2 \int_{1/2}^1 h^2(t)\, dt + 1/2 \int_0^{1/2} h^2(1-t)\, dt$$

$$= \int_{1/2}^1 h^2(t)\, dt.$$

Example 3.5.4: The Wilcoxon scores $a(i) = i \quad (1 \le i \le n)$ correspond to the score function $h(t)=t$. Transforming the one-sample Wilcoxon statistic as in Example 3.5.2,

$$N^{1/2}((N(n+1))^{-1}T - 1/2 \int g(1/2(F(t)+1-F(-t)))dF(t))$$

converges to $N(0,\sigma^2(g))$ in the D_2-metric, where $T=\sum R_i^+ \text{sign } X_i$, $g(t) = 2t-1$ and by (3.5.17) (if F is symmetric)

$$\sigma^2(g) = \int_{1/2}^1 (2t-1)^2\, dt = 1/6.$$

Exercise: Check the details.

<u>Theorem 3.5.4:</u> Let h be an absolutely continuous score function satisfying $|h''(t)| \leq K\,(t(1-t))^{-(5/2)+\delta}$ for some $K > 0$ and $\delta > 0$. Define $a_n(i) = E\,h(U_{(i)})$ where $U_{(i)}$ denotes the order statistic of n i.i.d. uniformly distributed random variables. Then

$$\lim_{n\to\infty} E\left(n^{-1/2}\left(\sum_{i=1}^{n} a_n(R_i^+)\operatorname{sign} X_i - \sum_{i=1}^{n} h\left(\frac{R_i^+}{n+1}\right)\operatorname{sign} X_i\right)\right)^2 = 0.$$

<u>Proof:</u> Of course, we may assume that h is increasing on $[1/2,1)$. Since

$$\left|\sum_{i=1}^{n}\left(a_n(R_i^+)-h\left(\frac{R_i^+}{n+1}\right)\right)\operatorname{sign} X_i\right| \leq \sum_{i=1}^{n} \left|Eh(U_{(i)}) - h\left(\frac{i}{n+1}\right)\right|$$

it suffices to prove

$$\lim_{n\to\infty} n^{-1/2} \sum_{1\leq i\leq n} \left|Eh(U_{(i)}) - h\left(\frac{i}{n+1}\right)\right| = 0.$$

But this is the same as (3.4.17) and has been proven for Theorem 3.4.4.

<u>Example 3.5.6:</u> Assume that the scores

$$a(i) = E\left(-(f'/f)\,F^{-1}(1/2(1+U_{(i)}))\right)$$

are given as in (3.5.6), where $(f'/f)\circ F^{-1} \in \mathcal{H}$ is absolutely continuous and has finite Fisher information

$$I(F) = \int \frac{f'(s)^2}{f(s)}\,ds.$$

By Theorems 3.5.3 and 3.5.4 (assuming 3.5.4 can be applied)

$$N^{1/2}\left(N^{-1}\sum_{i=1}^{n} a(R_i^+)\operatorname{sign} X_i - 1/2\int g(1/2(1+F(t)-F(-t)))dF(t)\right)$$

is asymptotically normal $\mathcal{N}(0,\sigma^2(g))$ (in the D_2-metric) where $g(t) = -(f'/f)\circ F^{-1}(t)$ if $t \geq 1/2$ and $g(t) = (f'/f)\circ F^{-1}(1-t)$ if $t < 1/2$. Here

$$\sigma^2(g) = \int_{1/2}^{1}\left((f'/f)\circ F^{-1}(t)\right)^2 dt = \int_{F^{-1}(1/2)}^{\infty} \frac{f'^2(s)}{f(s)}\,ds,$$

and if F is symmetric, then $\sigma^2(g) = 1/2\,I(F)$.

We conclude this section with a brief discussion of R-estimators. We start with the following

<u>Proposition 3.5.1:</u> Let $\{F_\vartheta : \vartheta \in \mathbb{R}\}$ be a location model. Assume that $T: \mathbb{R}^n \longrightarrow \mathbb{R}$ is a statistic satisfying

(3.5.18) For all $x = (x_1,\ldots,x_n) \in \mathbb{R}^n$ the function

$$a \longrightarrow T((x_1+a,\ldots,x_n+a))$$

is non-decreasing and has positive and negative values.

(3.5.19) The distribution of T under F_o^n is symmetric about 0.

Define $\hat{\vartheta} = 1/2\ (a^*+a_*)$, where

$$a^*(x) = \inf\ \{a : T((x_1-a,\ldots,x_n-a)) < 0\}$$

and

$$a_*(x) = \sup\ \{a : T((x_1-a,\ldots,x_n-a)) > 0\} \quad (x \in \mathbb{R}^n).$$

Then

(3.5.20) $-\infty < a_* \leq a^* < \infty$ (this holds already if (3.5.18) is satisfied),

(3.5.21) $F_\vartheta^n(\hat{\vartheta} \leq \vartheta) \geq 1/2\ (1 - F_o^n(T=0))$

(3.5.22) $F_\vartheta^n(\hat{\vartheta} \geq \vartheta) \geq 1/2\ (1 - F_o^n(T=0))$

and

(3.5.23) If F is continuous, then the distributions of a^* and a_* have no atoms.

<u>Proof:</u> (3.5.20) follows immediately from (3.5.18). Since (3.5.22) is similar to (3.5.21) , we only prove (3.5.21). For each $a \in \mathbb{R}$

$$F_a^n(\{x : T((x_1-a,\ldots,x_n-a)) \geq 0\}) = F_o^n(\{x : T(x) \geq 0\}) = $$
$$= 1/2 + 1/2\ F_o^n(T=0),$$

since $F_a(x) = F_o(x-a)$ and since T has a symmetric distribution with respect to F_o^n by (3.5.19). For $x \in \mathbb{R}^n$ consider the confidence interval $S(x) = \{a \in \mathbb{R} : T((x_1-a,\ldots,x_n-a)) < 0\}$.

Then

$$F_\vartheta^n(\{x : \vartheta \in S(x)\}) = F_\vartheta^n(\{x : T((x_1-\vartheta,\ldots,x_n-\vartheta)) < 0\}) =$$
$$= 1/2 - 1/2\ F_o^n(T=0).$$

By definition, $\vartheta \geq a^*(x)$ for $\vartheta \in S(x)$ and therefore $S(x) \subset [a^*(x),\infty)$ and

$$F_\vartheta^n(\hat{\vartheta} \leq \vartheta) \geq F_\vartheta^n(\{x : a^*(x) \leq \vartheta\}) \geq F_\vartheta^n(\{x : \vartheta \in S(x)\}) =$$
$$= 1/2 - 1/2\ F_o^n(T=0).$$

In order to prove (3.5.23) define the non-decreasing function $u_x(z) = T((z,x_2-x_1+z,\ldots,x_n-x_1+z))$ where $x=(x_1,\ldots,x_n) \in \mathbb{R}^n$ is fixed. Now

$$a^*(x) = \inf\ \{a : T((-(a-x_1),x_2-x_1-(a-x_1),\ldots,x_n-x_1-(a-x_1)) < 0\}$$

$$= \inf \{ b : T((-b, x_2-x_1-b, \ldots, x_n-x_1-b)) < 0 \} + x_1 =$$
$$= - \sup \{ b : u_x(b) < 0 \} + x_1,$$

and

$$F^n_\vartheta(a^*=c) = F^n_\vartheta(x_1=c+ \sup \{b : u_x(b) < 0 \}).$$

On a straight line G in $\mathbb{R}^n$ defined by $x_j-x_1 \equiv$ const. $(2 \leq j \leq n)$, the functions u_x are all equal and hence $| G \cap \{a^*=c\}| = 1$. Since F is continuous, F^n_ϑ has no atoms and its conditional measure on any line G as above has no atoms. Therefore, conditioning on lines G, $F^n_\vartheta(a^*=c \mid G) = 0$ and (3.5.23) follows from Fubini's theorem.

<u>Theorem 3.5.5:</u> Let $\{ F_\vartheta : \vartheta \in \mathbb{R}\}$ be a location model where $F = F_0$ is continuous, and let $\{T_n : n \geq 1\}$ be a sequence of statistics satisfying (3.5.18) and (3.5.19). Suppose that there exist $A \neq 0$, $\mu_n(u) \in \mathbb{R}$ and $\sigma_n > 0$ $(n \geq 1,\ u \in \mathbb{R})$ such that $\mu_n(0) = 0$, $\lim \sigma_n^{-1} \mu_n(u) = Au$ for all $u \in \mathbb{R}$ and $\sigma_n^{-1}(T_n - \mu_n(u)) \longrightarrow \mathcal{N}(0,1)$ weakly under $F_{un^{-1/2}}$ for all $u \in \mathbb{R}$. Then

$$(3.5.24) \qquad A\, n^{1/2} (\hat{\vartheta}_n - \vartheta) \longrightarrow \mathcal{N}(0,1)$$

weakly under F_ϑ, where $\hat{\vartheta}_n$ is defined in Proposition 3.5.1 and corresponds to T_n.

<u>Proof:</u> By definition and Proposition 3.5.1,(3.5.23), we have F^n_ϑ a.s.

$$\{\hat{\vartheta}_n < u\} \subset \{a_{n*} < u\} \subset \{T_n((x_1-u, \ldots, x_n-u)) \leq 0\}$$

and

$$\{\hat{\vartheta}_n < u\} \supset \{a^*_n \leq u\} \supset \{T_n((x_1-u, \ldots, x_n-u)) < 0\}.$$

Therefore it suffices to show that

$$\lim_{n\to\infty} F^n_\vartheta(\{ x : T_n((x_1-A^{-1}un^{-1/2}-\vartheta, \ldots, x_n-A^{-1}un^{-1/2}-\vartheta)) \leq 0\})$$
$$= (2\pi)^{-1/2} \int_{-\infty}^{u} \exp(-t^2/2)\, dt.$$

But the probabilities on the left-hand side are given by

$$F^n_{-A^{-1}un^{-1/2}}(\{ x : T_n(x) \leq 0 \}) =$$
$$= F^n_{-A^{-1}un^{-1/2}}(\{ x : \sigma_n^{-1}(T_n(x) - \mu_n(-\tfrac{u}{A})) \leq -\sigma_n^{-1}\mu_n(-\tfrac{u}{A})\})$$
$$\longrightarrow (2\pi)^{-1/2} \int_{-\infty}^{u} \exp(-t^2/2)\, dt,$$

since $\lim -\sigma_n^{-1}\mu_n(-u/A) = u$.

Proposition 3.5.1 and Theorem 3.5.5 especially apply to certain signed rank statistics.

Lemma 3.5.3: Let $T = \sum_{1\leq i\leq n} a(R_i^+) \operatorname{sign} X_i$ be a signed rank statistic. If for each $i = 1,\dots,n-1$ $a(i) \leq a(i+1)$, then (3.5.18) holds. Moreover, if the distribution F of X_1 is symmetric about 0, then (3.5.19) is true for F^n.

Proof: Write $T = \sum_{1\leq i\leq n} b(R_i)$ with scores

$$b(i) = \begin{cases} a(i-n) & \text{if } i > n \\ -a(n+1-i) & \text{if } i \leq n \end{cases} \qquad (1 \leq i \leq 2n),$$

where R_i denotes the rank of X_i among $\{X_j, -X_j : 1 \leq j \leq n\}$. If $a \geq 0$, then the rank of X_i+a among $\{X_j+a, -X_j-a : 1\leq j\leq n\}$ is at least R_i. Since $b(i) \leq b(i+1)$ it follows that $T((X_1+a,\dots,X_n+a)) \geq T(X)$ and (3.5.18) .

Next, let F be symmetric. Then $(-X_1,\dots,-X_n)$ has the same distribution as $(X_1,\dots,X_n)$. Obviously, this implies (3.5.19).

Definition 3.5.2: Let $T = \sum_{1\leq i\leq n} a(r_i^+) \operatorname{sign} x_i$ be a signed rank statistic with increasing scores $a(i)$. Define a^* and a_* as in Proposition 3.5.1. Then (3.5.20) holds by Lemma 3.5.3, since (3.5.18) implies (3.5.20). The statistic

$$(3.5.25) \qquad \hat{\vartheta}(x) = 1/2\ (\ a^*(x) + a_*(x)) \qquad (x\in\mathbb{R}^n)$$

is called an *R-estimator*.

Theorem 3.5.6: Let $\{F_\vartheta : \vartheta\in\mathbb{R}\}$ be a location model with symmetric, absolutely continuous d.f. $F = F_0$, and let $T_n(x) = \sum_{1\leq i\leq n} a(r_i^+) \operatorname{sign} x_i$ be a sequence of signed rank statistics with scores satisfying $a(i) = h(\frac{i}{n+1})$ where h is absolutely continuous and increasing with $h(0) = 0$. Define $g(t) = h(2t-1)$ if $t \geq 1/2$ and $g(t) = -h(1-2t)$ if $t < 1/2$. Assume that $g \in \mathcal{H}$ and that the function

$$t \longrightarrow g((1/2)(1 + F(x) - F(-x-t)))$$

is differentiable in some neighbourhood of 0 and has an $L_1(F)$-bounded derivative. If $\sigma^2 = \int_{1/2\ \leq t\leq\ 1} h^2(2t-1)\, dt$ and if f denotes the density function of F, then

$$(3.5.26) \qquad 2^{-1/2}\, \sigma^{-1} \left[\int g'(F(t))\ f^2(t)\ dt\right] n^{1/2}(\hat{\vartheta}_n-\vartheta) \longrightarrow \mathcal{N}(0,1)$$

weakly with respect to F_ϑ. (3.5.26) still holds if the scores are given by $a(i) = Eh(U_{(i)})$ and if h satisfies in addition the assumptions in Theorem 3.5.4.

Proof: We shall show the assumptions of Theorem 3.5.5. Then (3.5.26) follows immediately. From Theorem 3.5.3 (together with a reduction as in Example 3.5.4) we conclude that

$$N^{1/2}\left(N^{-1}T_n - (1/2)\int_{-\infty}^{\infty} g((1/2)(1+F_\vartheta(t)-F_\vartheta(-t)))\, dF_\vartheta(t)\right)$$

converges weakly to $\mathcal{N}(0,\sigma_\vartheta^2(g))$, and uniformly in F_ϑ, where $\sigma_\vartheta^2(g)$ is defined by (3.5.15) with $C_{iN} = 1$ if $i\leq n$ and with $C_{iN} = 0$ if $i>n$ (and therefore is independent of n because $H_N(t) = 1/2\ (1+F_\vartheta(t)-F_\vartheta(-t))$ and $\varphi(t) = 1/2\ (dF_\vartheta/dH_N)(t)$.)

We first claim that

$$\text{(3.5.27)}\qquad \lim_{\vartheta\to 0} \sigma_\vartheta^2(g) = \sigma_0^2(g) = \sigma^2(g).$$

$\sigma_\vartheta^2(g)$ is the variance of $W_N(g)$ defined in (3.5.14). Let us first assume that g' exists and is bounded and continuous. In order to prove (3.5.27) in this case it suffices to show that

$$\text{(3.5.28)}\qquad N\,E\left(\int_{-\infty}^{\infty} g(H_N(t))\, d(\hat{F}_N-F_N)(t)\right)^2$$

and

$$\text{(3.5.29)}\qquad N\,E\left(\int_{-\infty}^{\infty} (\hat{H}_N(t)-H_N(t))\,\frac{dF_N}{dH_N}(t)\, dg(H_N(t))\right)^2$$

(which are independent of N) are continuous in $\vartheta = 0$.

Let $X_1,\dots,X_n$ be i.i.d. F_0-distributed. Then $X_1+\vartheta,\dots,X_n+\vartheta$ are i.i.d. F_ϑ-distributed and therefore (3.5.28) is

$$1/2\ \operatorname{Var}_F\, g((1/2)(1 + F(X_1) - F(-X_1-2\vartheta))).$$

Since g is bounded and since the integrand converges pointwise to $g(F(X_1))$, (3.5.28) is continuous at $\vartheta = 0$ by Lebesgue's dominated convergence theorem.

In order to prove the continuity for (3.5.29) write this expression in the form

$$N\,E\left(\int_{-\infty}^{\infty} (\hat{H}_N(t)-H_N(t))\, g'(H_N(t))\, dF_N(t)\right)^2 =$$

$$= N\,E\left(1/2\int_{-\infty}^{\infty} (\hat{H}_N(t+\vartheta)-H_N(t+\vartheta))\, g'(1/2\ (1+F(t)-F(-t-2\vartheta)))\, dF(t)\right)^2.$$

This is continuous at $\vartheta = 0$, since for all s and t

$$N\,E\,(\hat{H}_N(t+\vartheta)-H_N(t+\vartheta))(\hat{H}_N(s+\vartheta)-H_N(s+\vartheta)) =$$
$$= 1/2\,E\,\big(1_{\{X_1\leq t\}}+1_{\{-X_1\leq t+2\vartheta\}}-(1+F(t)-F(-t-2\vartheta))\big)$$
$$\big(1_{\{X_1\leq s\}}+1_{\{-X_1\leq s+2\vartheta\}}-(1+F(s)-F(-s-2\vartheta))\big)$$

is continuous at $\vartheta = 0$. (Exercise)

We have shown that (3.5.27) holds for all $g \in C_{1b}$.

In view of Theorem 3.5.2 we have for any two score functions $g_1, g_2 \in H$

$$|ET_N(g_1)^2 - ET_N(g_2)^2| \leq K\,\|g_1 - g_2\|^2$$

uniformly over all continuous distributions. Also from the D_2-convergence in Theorem 3.5.3 it follows for absolutely continuous proper score functions g_1 that

$$\lim_{N\to\infty} |ET_N(g_1)^2 - \sigma^2(g_1)| = 0$$

uniformly over all continuous d.f. It follows that for absolutely continuous score functions $g_1, g_2 \in H$

$$|\sigma^2(g_1) - \sigma^2(g_2)| \leq K\,\|g_1 - g_2\|^2$$

uniformly over all continuous d.f. Therefore $\sigma^2_\vartheta(g)$ is continuous at $\vartheta = 0$ for arbitrary absolutely continuous $g \in H$, proving (3.5.27).

Define

$$\mu_n(u) = n \int g(1/2\,(1 + F(t-un^{-1/2}) - F(-t-un^{-1/2})))\,dF(t-un^{-1/2})$$

for $u \in \mathbb{R}$. By (3.5.27) and Theorem 3.5.3

$$\sigma^{-1}(g)\,N^{-1/2}\,(T_n - \mu_n(u)) \longrightarrow N(0,1) \quad (\text{rel. to } F_{un^{-1/2}})$$

in the D_2-metric for all $u \in \mathbb{R}$.

Furthermore, note that

$$\mu_n(0) = n \int g(t)\,dt = n\Big(\int_{1/2}^{1} h(2t-1)\,dt - \int_0^{1/2} h(1-2t)\,dt\Big) = 0$$

and that

$$\lim_{n\to\infty} N^{-1/2}\,\sigma^{-1}(g)\,\mu_n(u) =$$

$$= 2^{-1/2}\,\sigma^{-1}(g)\,\lim_{n\to\infty} n^{1/2} \int g(1/2\,(1+F(t)-F(-t-2un^{-1/2})))\,dF(t) =$$

$$= 2^{1/2}\, u\sigma^{-1}(g) \lim_{n\to\infty} \int \frac{g(1/2\ (1+F(t)-F(-t-2un^{-1/2})))-g(F(t))}{2un^{-1/2}}\, dF(t) =$$

$$= 2^{-1/2}\, u\, \sigma^{-1}(g) \int g'(F(t))\, f^2(t)\, dt.$$

From Example 3.5.4 it is clear that $\sigma^2(g) = \sigma^2$ and therefore (3.5.26) is proven.

Example 3.5.7: The R-estimator $\hat{\vartheta}$ corresponding to the Wilcoxon one-sample statistic is called the *Hodges-Lehmann estimator*. We shall see that $\hat{\vartheta}$ estimates the median of the empirical measure

$$\nu_x = (2n)^{-1} \sum_{i=1}^{n} \left(\sum_{j=1}^{i-1} \varepsilon_{1/2\ (x_{(i)}+x_{(j)})} + \varepsilon_{x_{(i)}} \right).$$

This follows from

$$a^*(x) = \inf\ \{a : \sum_{i=1}^{n} r_i^+(x_i-a)\ \mathrm{sign}\ (x_i-a) < 0\ \} =$$

$$= \inf\ \{a : \sum_{\substack{i=1\\ x_i>a}}^{n} \sum_{j=1}^{n} 1_{\{x_i+x_j>2a\}}\, 1_{\{x_i\geq x_j\}} - \sum_{\substack{i=1\\ x_i\leq a}}^{n} \sum_{j=1}^{n} 1_{\{x_i<x_j\}}\, 1_{\{x_i+x_j\leq 2a\}} < 0\} =$$

$$= \inf\ \{a : \sum_{i=1}^{n} \left(\sum_{j=1}^{i-1} 1_{\{x_{(i)}+x_{(j)}>2a\}} + 1_{\{x_{(i)}>a\}} \right) - n < 0\} =$$

$$= \inf\ \{a : \nu_x((a,\infty)) < 1/2\},$$

and similarly

$$a_*(x) = \sup\ \{a : \nu_x((a,\infty)) > 1/2\}.$$

Instead of taking the median of ν_x for the Hodges-Lehmann estimator, the median of $\sum_{1\leq i\leq n} \varepsilon_{1/2\ (x_{(i)}+x_{(n+1-j)})}$ defines the *Bickel-Hodges estimator* which is not an R-estimator. Certainly, normal scores $a(i) = E\ \Phi^{-1}(1/2 + 1/2\ U_{(i)})$ or $a(i) = \Phi^{-1}(1/2 + i/2(n+1))$ also define R-estimators.

If all the requirements of Theorem 3.5.6 about the distributions are fulfilled, then (for the Hodges-Lehmann estimator $\hat{\vartheta}_n$) $n^{1/2}\, \hat{\vartheta}_n$ will be asymptotically normal under F with zero expectation and variance

$$2\,\sigma^2\Big(\int g'(F(t))\,f^2(t)\,dt\Big)^{-2} = \frac{1}{3}\Big(\int f^2(t)\,dt\Big)^{-2}.$$

Example 3.5.8: Let F be a symmetric, continuous d.f. which has a density f with the following properties: log f is concave and three times differentiable such that (log f)'' has an integrable bound. Assume furthermore that the score function h defined by

$$h(t) = -\,(f'/f)\circ F^{-1}(1/2 + 1/2\,t)$$

satisfies

$$|h''(t)| \leq K\,(t(1-t))^{-5/2\,+\,\delta}$$

for some $\delta > 0$ and $K > 0$. Then $h \in \mathcal{H}$ and h is increasing since

$$h'(t) = -1/2\,(\log f)''(F^{-1}(1/2 + 1/2\,t))(f(F^{-1}(\tfrac{1}{2}+\tfrac{t}{2})))^{-1} \geq 0.$$

If

$$g(t) = \begin{cases} h(2t-1) = -(f'/f)\circ F^{-1}(t) & (t \geq 1/2) \\ -h(1-2t) = (f'/f)\circ F^{-1}(1-t) & (t < 1/2) \end{cases}$$

is defined as before, then $(g\circ F)' = (\log f)''$. Therefore Theorem 3.5.6 can be applied to the signed rank statistics $T_n(x) = \sum a_n(r_i^+)\,\text{sign}\,x_i$ where

$$a_n(i) = Eh(U_{(i)}) \qquad (\,1 \leq i \leq n\,).$$

($U_{(1)} < U_{(2)} < \ldots < U_{(n)}$ denotes again the order statistic of n i.i.d. uniformly distributed random variables.) The corresponding R-estimators $\hat{\vartheta}_n$ are asymptotically normal under F with zero expectation and variance τ^2 given by

$$\tau^2 = 2\,\sigma^2(g)\,\Big(\int g'(F(t))\,f^2(t)\,dt\Big)^{-2}.$$

We have

$$\sigma^2(g) = \int_{1/2}^{1}\Big((f'/f)\circ F^{-1}(t)\Big)^2\,dt =$$

$$= \int_0^{\infty} (f'(t)/f(t))^2\,dF(t) = I(F)/2$$

where I(F) denotes the Fisher information (cf. Example 3.5.6), and

$$\int g'(F(t))\,f^2(t)\,dt = \int (g\circ F)'(t)\,dF(t) =$$

$$= - \int \frac{f''(t)f(t) - f'^2(t)}{f^2(t)} \, f(t) \, dt$$

$$= \int f'^2(t)/f(t) \, dt = I(F).$$

Therefore $\tau^2 = 1/I(F).$

6. Linear combinations of a function of the order statistic and L-estimators

Definition 3.6.1: A measurable map $T : \mathbb{R}^n \longrightarrow \mathbb{R}$ is called a *linear combination of a function of the order statistic* if there exist a measurable function $f : \mathbb{R} \longrightarrow \mathbb{R}$ and constants $c_i \in \mathbb{R}$ such that

$$(3.6.1) \qquad T(x) = \sum_{i=1}^{n} c_i \, f(x_{(i)}) \qquad (x=(x_1,\ldots,x_n) \in \mathbb{R}^n)$$

where $x_{(1)} \leq x_{(2)} \leq \ldots \leq x_{(n)}$ denote the ordered coordinates or x. If f is the identity, then T is called a *linear combination of the order statistic* or *L-estimator* (if T is used to estimate a location).

At a first glance it is not clear how these statistics fit into the general concept of this chapter. However, setting $a(i) = c_i \quad (1 \leq i \leq n)$ we may write

$$(3.6.2) \qquad T(x) = \sum_{i=1}^{n} f(x_i) \, a(r_i)$$

and thus we may consider T as a simple linear rank statistic with scores a(i) and generalized random regression constants $C_i = f(x_i)$. In this section we shall prove the asymptotic normality of T if $\{X_i : i \geq 1\}$ are i.i.d., extending the method of proof in section 4 once more. Its application to asymptotic efficiency will be given in the next chapter. We begin with a few more examples.

Example 3.6.1: The mean $1/n \sum X_i$ may be considered also as a linear combination of the order statistic. Very similar is *Gini's mean difference*

$$\frac{2}{n(n-1)} \sum_{1 \leq i < j \leq n} |x_i - x_j| = (n(n-1))^{-1} \sum_{i,j=1}^{n} |x_i - x_j| =$$

$$= (n(n-1))^{-1} \sum_{i,j=1}^{n} |x_{(i)} - x_{(j)}| =$$

$$= 2(n(n-1))^{-1} \sum_{i=1}^{n} \sum_{j=i+1}^{n} (x_{(j)} - x_{(i)}) =$$

$$= 2(n(n-1))^{-1} \sum_{i=1}^{n} (2i-n-1)\, x_{(i)}.$$

Observe that Gini's mean difference also may be investigated as a U-statistic.

Example 3.6.2: The sample p-th quantile (fractile) of a continuous d.f. F was already introduced in Example 2.1.7 as a differentiable statistical functional. It is given by $F_n^{-1}(p)$ where F_n denotes the empirical d.f. of n i.i.d. F-distributed random variables. Also it may be treated as a linear combination of the order statistic because of

$$F_n^{-1}(p) = \inf\{\, x : F_n(x) \geq p\} = X_{(i)}$$

where $(i-1)/n < p \leq i/n$.

Defining $c_j = 1$ if $(j-1)/n < p \leq j/n$ and $c_j = 0$ otherwise then

$$F_n^{-1}(p) = \sum_{1 \leq j \leq n} c_j\, X_{(j)}\ .$$

Example 3.6.3: Let $0 \leq \alpha \leq 1/2$. The α-trimmed mean (cf. Example 2.1.8) is defined by

$$\frac{1}{n-2[n\alpha]} \sum_{i=[n\alpha]+1}^{n-[n\alpha]} X_{(i)}$$

and the α-*winsorized mean* by

$$n^{-1}\left([n\alpha]\, X_{([n\alpha]+1)} + [n\alpha]\, X_{(n-[n\alpha])} + \sum_{i=[n\alpha]+1}^{n-[n\alpha]-1} X_{(i)} \right).$$

Let $T = \sum_{i=1}^{n} c_i\, f(X_{(i)})$ be a linear combination of a function of the order statistic and let $c_i = h(i/(n+1))$ be given by a score function $h : (0,1) \longrightarrow \mathbb{R}$. It follows from (3.6.2) that T can be written as

$$T = n \int_{-\infty}^{\infty} h\left(\tfrac{n}{n+1}\hat{H}_n(t)\right) f(t)\, d\hat{H}_n(t) \tag{3.6.3}$$

where - as before - $\hat{H}_n(t) = n^{-1} \sum_{1 \leq i \leq n} 1_{\{X_i \leq t\}}$ denotes the empirical d.f. of $X_1,\dots,X_n$.

In a very first step we extend the notion of a proper score function. Let $\delta \geq 0$ and $\eta = \delta\,(4+2\delta)^{-1}$. $\mathcal{H}_\delta$ denotes the Banach space of all right continuous functions $h : (0,1) \longrightarrow \mathbb{R}$ of bounded variation on each compact subset of $(0,1)$ satisfying $h(1/2) = 0$ and

$$(3.6.4) \qquad \|h\|_\delta = \int_0^1 (t(1-t))^{-1/2-\eta}(|h_1(t)|+|h_2(t)|)\,dt<\infty$$

where $h = h_1 - h_2$ is the Hahn-Jordan decomposition of h according to Lemma 3.3.2. The set C_{2b} of functions with bounded second derivative is dense in the subspace of all absolutely continuous functions in $\mathcal{H}_\delta$. Since $\mathcal{H} = \mathcal{H}_0$ the present definition generalizes Definition 3.3.1 . For a proof of these statements use the following generalization of Lemma 3.3.4 and proceed as in the proof of Lemma 3.3.5 (Exercise).

<u>Lemma 3.6.1:</u> There exists a constant C, depending on δ, such that for $h \in \mathcal{H}_\delta$ the following holds:

$$(3.6.5) \qquad \lim_{z\to\{0,1\}} |h(z)|\,(z(1-z))^{1/2-\eta} = 0,$$

$$(3.3.6) \qquad \sup_{0<z<1} |h(z)|\,(z(1-z))^{1/2-\eta} \leq \|h\|_\delta ,$$

$$(3.6.7) \qquad \int_0^1 (z(1-z))^{1/2-\eta}|dh|(z) \leq \|h\|_\delta \leq$$

$$\leq C \int_0^1 (z(1-z))^{1/2-\eta}\,|dh|(z)$$

where $|dh|$ denotes the total variation measure of dh, and

$$(3.6.8) \qquad \int_0^1 |h(z)|^{2+\delta}\,dz \leq \|h\|_\delta^{2+\delta} .$$

The proof of this lemma is the same as that one of Lemma 3.3.4 with the obvious modifications. We leave it as an exercise. The approximation theorem for linear combinations of a function of the order statistic, however, needs a proof. It has the following form:

<u>Theorem 3.6.1:</u> Let $\alpha,\beta \geq 0$ satisfy $\alpha\beta \geq 4$. There exists a constant $K > 0$ such that for every sequence $X_1,\dots,X_n$ of i.i.d. random variables with continuous d.f. F and for every

pair of functions f and g with $f \circ F^{-1} \in H_\alpha$ and $g \in H_\beta$ we have

$$(3.6.9) \quad n\, E\left(\int_{-\infty}^{\infty} f(t)\, g(\tfrac{n}{n+1}\hat{H}_n(t))\, d\hat{H}_n(t) - \right.$$
$$\left. - \int_{-\infty}^{\infty} f(t)\, g(F(t))\, dF(t) \right)^2 \leq K \|f \circ F^{-1}\|_\alpha^2\, \|g\|_\beta^2 .$$

Proof: We shall estimate separately in L_2

$$(3.6.10) \quad n^{1/2} \int_{-\infty}^{\infty} f(t)\, (g(\tfrac{n}{n+1}\hat{H}_n(t)) - g(\tfrac{n}{n+1}F(t)))\, d\hat{H}_n(t)$$

$$(3.6.11) \quad n^{1/2} \int_{-\infty}^{\infty} f(t)\, g(\tfrac{n}{n+1}F(t))\, d(\hat{H}_n - F)(t)$$

and

$$(3.6.12) \quad n^{1/2} \int_{-\infty}^{\infty} f(t)\, (g(\tfrac{n}{n+1}F(t)) - g(F(t)))\, dF(t) .$$

We may assume that f and g are increasing since the norms $\| \ \|_\alpha$ and $\| \ \|_\beta$ respect the Hahn-Jordan decomposition.

Proof of (3.6.10): Define

$$A_1 = \{(z,s,t) : F^{-1}(1/2) < s \leq t;\ \tfrac{n}{n+1}F(t) < z \leq \tfrac{n}{n+1}\hat{H}_n(t)\} =$$
$$= \{(z,s,t) : F^{-1}(1/2) < s \leq t;\ \hat{H}_n^{-1}(\tfrac{n+1}{n}z) \leq t < F^{-1}(\tfrac{n+1}{n}z)\} ,$$

$$B_1 = \{(z,s,t) : F^{-1}(1/2) < s \leq t;\ \tfrac{n}{n+1}\hat{H}_n(t) < z \leq \tfrac{n}{n+1}F(t)\} =$$
$$= \{(z,s,t) : F^{-1}(1/2) < s \leq t;\ F^{-1}(\tfrac{n+1}{n}z) \leq t < \hat{H}_n^{-1}(\tfrac{n+1}{n}z)\} ,$$

$$A_2 = \{(z,s,t) : t < s \leq F^{-1}(1/2);\ \tfrac{n}{n+1}F(t) < z \leq \tfrac{n}{n+1}\hat{H}_n(t)\} =$$
$$= \{(z,s,t) : t < s \leq F^{-1}(1/2);\ \hat{H}_n^{-1}(\tfrac{n+1}{n}z) \leq t < F^{-1}(\tfrac{n+1}{n}z)\}$$

and

$$B_2 = \{(z,s,t) : t < s \leq F^{-1}(1/2);\ \tfrac{n}{n+1}\hat{H}_n(t) < z \leq \tfrac{n}{n+1}F(t)\} =$$
$$= \{(z,s,t) : t < s \leq F^{-1}(1/2);\ F^{-1}(\tfrac{n+1}{n}z) \leq t < \hat{H}_n^{-1}(\tfrac{n+1}{n}z)\} .$$

Let $\varphi = 1_{A_1} - 1_{B_1} + 1_{B_2} - 1_{A_2}$. Then we obtain

$$\int_{-\infty}^{\infty} f(t) \left(g(\tfrac{n}{n+1}\hat{H}_n(t)) - g(\tfrac{n}{n+1}F(t)) \right) d\hat{H}_n(t) =$$

$$= \iiint 1_{\{z \leq n/(n+1)\}} \left(1_{(F^{-1}(1/2),\infty)}(t)\, 1_{(F^{-1}(1/2),t]}(s) - \right.$$

$$- 1_{(-\infty, F^{-1}(1/2)]}(t)\, 1_{(t, F^{-1}(1/2)]}(s)\Big)$$

$$\Big(1_{\{(z,t): \frac{n}{n+1} F(t) < z \leq \frac{n}{n+1} \hat{H}_n(t)\}} - 1_{\{(z,t): \frac{n}{n+1} \hat{H}_n(t) < z \leq \frac{n}{n+1} F(t)\}} \Big)$$

$$dg(z)\, df(s)\, d\hat{H}_n(t) =$$

$$= \int_0^{n/n+1} \int_{-\infty}^{\infty} \int_{-\infty}^{\infty} \varphi(z,s,t)\, d\hat{H}_n(t)\, df(s)\, dg(z) .$$

Using Fubini's theorem and Cauchy-Schwarz' inequality it follows that

$$n\, E\Big(\int_{-\infty}^{\infty} f(t) \big(g(\tfrac{n}{n+1}\hat{H}_n(t)) - g(\tfrac{n}{n+1}F(t))\big)\, d\hat{H}_n(t) \Big)^2 =$$

$$= n \int_0^{n/n+1} \int_0^{n/n+1} \int_{-\infty}^{\infty} \int_{-\infty}^{\infty} E\Big(\int_{-\infty}^{\infty} \int_{-\infty}^{\infty} \varphi(z,s,t)\, \varphi(z',s',t')\, d\hat{H}_n(t)\, d\hat{H}_n(t') \Big)\, df(s)\, df(s')\, dg(z) dg(z')$$

$$\leq n \Big(\int_0^{n/n+1} \int_{-\infty}^{\infty} \Big\{ E\big(\int_{-\infty}^{\infty} |\varphi(z,s,t)|\, d\hat{H}_n(t)\big)^2 \Big\}^{1/2} df(s) dg(z) \Big)^2 .$$

Obviously, (3.6.10) is bounded by $K \| f \circ F^{-1} \|_\alpha^2 \| g \|_\beta^2$ if we can show that

$$(3.6.13) \qquad \psi(z,s) = n\, E\big(\int_{-\infty}^{\infty} |\varphi(z,s,t)|\, d\hat{H}_n(t)\big)^2$$

$$\leq \Lambda \min\big(z(1-z),\, F(s)(1-F(s)) \big)$$

holds for some constant $\Lambda > 0$ and all $s \in \mathbb{R}$ and $0 < z \leq \frac{n}{n+1}$. Indeed, since $\alpha\beta \geq 4$, $\alpha/(2+\alpha) + \beta/(2+\beta) \geq 1$ and consequently

$$\psi(z,s) \leq \Lambda\, (z(1-z))^{1-\beta/(2+\beta)}\, (F(s)(1-F(s)))^{1-\alpha/(2+\alpha)} .$$

Then (3.6.10) follows from Lemma 3.6.1 (3.6.7).

Unfortunately, in order to prove (3.6.13) we have to consider five different cases.

Case 1: $F^{-1}(1/2) < s \leq F^{-1}(\frac{n+1}{n} z)$, $F(s) \leq \frac{n}{n+1}$.

Since

$$\int_{-\infty}^{\infty} |\varphi(z,s,t)|\, d\hat{H}_n(t) \leq 2 \min(n^{-1}, z) + |\hat{H}_n(F^{-1}(\tfrac{n+1}{n} z)-) - \tfrac{n+1}{n} z| ,$$

$$\psi(z,s) \leq 4 \min (n^{-1}, nz^2) + 2 \frac{n+1}{n} z(1- \frac{n+1}{n}z) \leq$$

$$\leq 10 \min \{ z(1-z), F(s)(1-F(s)) \} .$$

Case 2: $F^{-1}(1/2) < s \leq F^{-1}(\frac{n+1}{n}z)$, $F(s) > \frac{n}{n+1}$.

In this case $\frac{n+1}{n}z > \frac{n}{n+1} \geq \frac{n-1}{n}$ and hence $\hat{H}_n(\hat{H}_n^{-1}(\frac{n+1}{n}z)-) = 1$.
Therefore

$$\int_{-\infty}^{\infty} |\varphi(z,s,t)| \, d\hat{H}_n(t) \leq |1 - \hat{H}_n(F^{-1}(\frac{n+1}{n}z)-)|$$

and $$\psi(z,s) \leq 2n(1 - \frac{n+1}{n}z)^2 + 2 \frac{n+1}{n}z(1- \frac{n+1}{n}z) \leq$$

$$\leq 4 (1 - \frac{n+1}{n}z) \leq 16 \min \{ z(1-z), F(s)(1-F(s))\} .$$

Case 3: $F^{-1}(1/2) < s$, $F^{-1}(\frac{n+1}{n}z) < s$.

Here $A_1 = \emptyset$ and hence

$$\int_{-\infty}^{\infty} |\varphi(z,s,t)| \, d\hat{H}_n(t) \leq \max \{ 0, \hat{H}_n(\hat{H}_n^{-1}(\frac{n+1}{n}z)-) - \hat{H}_n(s-)\} \leq$$

$$\leq 1 - F(s) + |\hat{H}_n(s-) - F(s)|$$

and

$$\int_{-\infty}^{\infty} |\varphi(z,s,t)| \, d\hat{H}_n(t) \leq 2 \min (n^{-1}, z) + $$
$$+ |\hat{H}_n(F^{-1}(\frac{n+1}{n}z)-) - \frac{n+1}{n}z| .$$

Consequently

$$\psi(z,s) \leq 10 \min \{ z(1-z), F(s)(1-F(s))\} .$$

Case 4: $s \leq F^{-1}(1/2)$, $s \geq F^{-1}(\frac{n+1}{n}z)$.

Clearly

$$\int_{-\infty}^{\infty} |\varphi(z,s,t)| \, d\hat{H}_n(t) \leq 2 \min (n^{-1}, z) +$$
$$+ |\hat{H}_n(F^{-1}(\frac{n+1}{n}z)-) - \frac{n+1}{n}z|$$

and (3.6.13) follows.

Case 5: $s \leq F^{-1/2}(1/2)$, $s < F^{-1}(\frac{n+1}{n}z)$.

We have

$$\int_{-\infty}^{\infty} |\varphi(z,s,t)| \, d\hat{H}_n(t) \leq$$

$$\leq \min \{\hat{H}_n(F^{-1}(\frac{n+1}{n}z)-)-\hat{H}_n(\hat{H}_n^{-1}(\frac{n+1}{n}z)-);\hat{H}_n(s-)-\hat{H}_n(\hat{H}_n^{-1}(F(s))-)\}$$

and, similarly as before,

$$\psi(z,s) \leq 10 \min \{ z(1-z), F(s)(1-F(s))\} .$$

Proof of (3.6.11):

Using (3.6.6) it follows that

$$E\left(n^{-1/2}\sum_{i=1}^{n} f(X_i)g(\tfrac{n}{n+1}F(X_i)) - Ef(X_i)g(\tfrac{n}{n+1}F(X_i))\right)^2$$

$$\leq \int_{-\infty}^{\infty} f^2(t)\ g^2(\tfrac{n}{n+1}F(t))\ dF(t) =$$

$$= \int_0^1 f^2(F^{-1}(t))\ g^2(\tfrac{n}{n+1}t)\ dt \leq$$

$$\leq \int_0^1 \frac{|f\circ F^{-1}(t)|\ \|f\circ F^{-1}\|_\alpha\ \|g\|_\beta^2}{(\frac{n}{n+1}t(1-\frac{n}{n+1}t))^{1-\beta/2+\beta}\ (t(1-t))^{1/2-\alpha/[4+2\alpha]}}\ dt \leq$$

$$\leq 2\ \|f\circ F^{-1}\|_\alpha\ \|g\|_\beta^2 \int_0^1 |f\circ F^{-1}(t)|\ (t(1-t))^{-1/2-\alpha/[4+2\alpha]}\ dt \leq$$

$$\leq 2\ \|f\circ F^{-1}\|_\alpha^2\ \|g\|_\beta^2\ ,$$

since $1/2 + \alpha/4+2\alpha \geq 3/2 - \beta/2+\beta - \alpha/4+2\alpha$.

Proof of (3.6.12):

Using Lemma 3.6.1 again it is not hard to see that

$$n^{1/2}\int_{-\infty}^{\infty} f(t)\ (g(F(t)) - g(\tfrac{n}{n+1}F(t)))\ dF(t) =$$

$$= n^{1/2}\int_0^1\int_{-\infty}^{\infty} f(t)\ 1_{\{\frac{n}{n+1}F(t)<z\leq F(t)\}}\ dF(t)\ dg(z) \leq$$

$$\leq n^{1/2}\ \|f\circ F^{-1}\|_\alpha \int_0^1\int_0^1 (t(1-t))^{-1/2+\alpha/[4+2\alpha]}\ 1_{\{z\leq t<\frac{n+1}{n}z\wedge 1\}}\ dt\ dg(z) \leq$$

$$\leq \text{const.}\ n^{1/2}\ \|f\circ F^{-1}\|_\alpha \int_0^1 n^{-1/2}(z(1-z))^{\alpha/[4+2\alpha]}\ dg(z) \leq$$

$$\leq K\ \|f\circ F^{-1}\|_\alpha\ \|g\|_\beta\ .$$

Theorem 3.6.2: Let $\alpha,\beta \geq 0$ satisfy $\alpha\beta \geq 4$, and let $X_1, X_2, \ldots$ be a sequence of i.i.d. random variables with continuous d.f. F. Assume that $f\circ F^{-1} \in \mathcal{H}_\alpha$ and $g \in \mathcal{H}_\beta$ are absolutely continuous. Then the corresponding linear combination of a function of the order statistic is asymptotically normal, i.e.

$$\sigma^{-1}n^{1/2}\left(\int_{-\infty}^{\infty} f(t)\ g(\tfrac{n}{n+1}\hat{H}_n(t))\ d\hat{H}_n(t) - \int_{-\infty}^{\infty} f\circ F^{-1}(t)g(t)\ dt\right)$$

converges to $N(0,1)$ with respect to the D_2-metric, where

(3.6.14) $\sigma^2 = 2 \iint_{s<t} s(1-t)\; g(t)\; g(s)\; df(F^{-1}(s))\; df(F^{-1}(t)).$

Proof: Assume first that g has a bounded second derivative. Taylor's theorem implies

$$g(\tfrac{n}{n+1}\hat{H}_n(t)) = g(F(t)) + g'(F(t))\; (\tfrac{n}{n+1}\hat{H}_n(t) - F(t)) + $$
$$+ \tfrac{1}{2}\, g''(\vartheta(\hat{H}_n(t)))\; (\tfrac{n}{n+1}\hat{H}_n(t) - F(t))^2,$$

and

$$n^{1/2}\left(\int_{-\infty}^{\infty} f(t)\; g(\tfrac{n}{n+1}\hat{H}_n(t))\; d\hat{H}_n(t) - \int_{-\infty}^{\infty} f(t)\; g(F(t))\; dF(t)\right)$$

$$= n^{1/2} \int_{-\infty}^{\infty} f(t)\; g(F(t))\; d(\hat{H}_n - F)(t) +$$

$$+ n^{1/2} \int_{-\infty}^{\infty} f(t)\; g'(F(t))\; (\hat{H}_n(t) - F(t))\; dF(t) +$$

$+ A + B + C$, where

$$A = n^{1/2} \int_{-\infty}^{\infty} f(t)\; g'(F(t))\; (\tfrac{n}{n+1}\hat{H}_n(t) - F(t))\; d(\hat{H}_n - F)(t),$$

$$B = -\frac{n^{1/2}}{n+1} \int_{-\infty}^{\infty} f(t)\; g'(F(t))\; \hat{H}_n(t)\; dF(t)$$

and

$$C = \tfrac{1}{2}\, n^{1/2} \int_{-\infty}^{\infty} f(t)\; g''(\vartheta(\hat{H}_n(t)))\; (\tfrac{n}{n+1}\hat{H}_n(t) - F(t))^2\; d\hat{H}_n(t)\;.$$

Using Lemma 3.3.6 it follows that

$$E\, A^2 \le c\, n^{-1} \int_0^1 f(F^{-1}(t))^2\; g'(t)^2\; dt \le \text{const.}\; n^{-1}\; \| f\circ F^{-1}\|_\alpha^2$$

$$E\, B^2 \le \text{const.}\; n^{-1}\; \| f\circ F^{-1}\|_\alpha^2 \quad \text{and}$$

$$E\, C^2 \le 1/4\; cn \sum_{l=0}^{2} n^{-4+l} \int_0^1 f(F^{-1}(t))^2\; (t(1-t))^l\; dt \le$$

$$\le \text{const.}\; n^{-1}\; \| f\circ F^{-1}\|_\alpha^2\;.$$

Define

$$W_n = n^{1/2}\left(\int_{-\infty}^{\infty} f(t)\; g(F(t))\; d(\hat{H}_n - F)(t) + \right.$$

$$\left. + \int_{-\infty}^{\infty} f(t)\; g'(F(t))\; (\hat{H}_n(t) - F(t))\; dF(t)\right) =$$

$$= -\, n^{1/2} \int_{-\infty}^{\infty} (\hat{H}_n(t) - F(t))\; g(F(t))\; df(t).$$

The last equality is obtained by partial integration observing that $\lim_{t \to \pm\infty} (\hat{H}_n(t)-F(t))\, f(t)\, g(F(t)) = 0$, the continuity of F, f and g and the right continuity of $\hat{H}_n$. Since W_n is the sum of n i.i.d. square integrable random variables,

$$\lim_{n \to \infty} D_2(\mathcal{L}(W_n), \mathcal{N}(0,\sigma^2)) = 0,$$

where $\sigma^2 = \text{Var } W_n$. In fact σ^2 does not depend on n since

$$\sigma^2 = n \iint E\Big((\hat{H}_n(t)-F(t))\, (\hat{H}_n(s)-F(s))\Big)\, g(F(t))\, g(F(s))\, df(t)\, df(s) =$$

$$= 2 \iint_{s<t} E\Big((1_{\{X_1 \leq t\}} - F(t))(1_{\{X_1 \leq s\}} - F(s))\Big)\, g(F(t))\, g(F(s))\, df(t)\, df(s) =$$

$$= 2 \iint_{s<t} s(1-t)\, g(t)\, g(s)\, df(F^{-1}(t))\, df(F^{-1}(s)) .$$

Since an arbitrarily chosen pair (f,g) as in the theorem can be approximated by pairs (f,g_m) where $g_m \in C_{2b}$, the theorem follows analoguously to the proof of Theorem 3.4.3, once it has been verified that σ^2 depends continuously on f and g. Let f and g be increasing. Then by (3.6.6) and (3.6.7) we have

$$\sigma^2 \leq 2 \Big(\int_0^1 (t(1-t))^{1/2}\, g(t)\, df(F^{-1}(t))\Big)^2 \leq$$

$$\leq 2\, \|g\|_\beta^2 \Big(\int_0^1 (t(1-t))^{\beta/[4+2\beta]}\, df(F^{-1}(t))\Big)^2 \leq$$

$$\leq 2\, \|g\|_\beta^2\, \|f \circ F^{-1}\|_\alpha^2 .$$

<u>Example 3.6.4:</u> For Gini's mean difference T_n in Example 3.6.1 $f(t) = t$ and $g(t) = 4t - 2$. Hence if $F^{-1} \in \mathcal{H}$,

$$n^{1/2} (T_n - 2 \int_0^1 F^{-1}(t)\, (2t-1)\, dt)$$

is asymptotically normal with variance

$$\sigma^2 = 8 \iint_{u<v} F(u)(1-F(v))(2F(v)-1)(2F(u)-1)\, du\, dv .$$

Notes on chapter 3:

The proof of Theorem 3.1.1 on the asymptotic distribution of permutation statistics is due to Hájek (1961). For a detailed discussion of the different conditions under which such a theorem were proved we refer to the book of Puri and Sen (1971). An excellent reference for rank tests and its asymptotic distribution theory under the hypothesis is the book of Hájek and Šidák (1967). Theorems 3.2.1 and 3.5.1 are taken from it. The notion of locally most powerfull tests was first used by Hoeffding (1950). The representation of rank statistics in Lemma 3.3.3 is used to derive the asymptotic normality of simple linear rank statistics using the Pyke-Shorack approach (1968). However, in sections 3-6 the results are derived by a method developed by Rösler and the author (1981,1982). Theorems 3.4.2,3.5.2,3.6.1,3.6.2, Lemmas 3.3.4,3.6.1 are taken from it. Lemma 3.3.6 is contained in Puri,Rösler and the author (1982). The results on the asymptotic normality in Theorems 3.4.3 and 3.5.3 follow from Hájek's work (1968). R-estimators were defined by Hodges and Lehmann (1963), especially Theorems 3.5.5 and 3.5.6 are due to them. Multivariate versions of rank statistics are discussed in Puri and Sen (1971); the presentation of this chapter carries over.

Chapter 4: Contiguity and efficiency

This chapter provides some basic idea of the concepts of efficiency and contiguity. The uniform convergence with respect to the distribution functions in Theorems 3.4.3, 3.5.3 and 3.6.2 may be seen as a typical consequence of contiguity: If for each $n \in \mathbb{N}$ the statistic T_n is considered under different distributions P_n, then the asymptotic normality still holds. Especially this is true when the sequence $P_n = (F_n)^n$ $(n \in \mathbb{N})$ is contiguous to the sequence F^n $(n \in \mathbb{N})$.
The computation of efficiency is mostly concerned with the determination of the variance of the limit distribution of the non-normalized statistics. In order to give the idea of efficiency we shall restrict to Pitman efficiency only. Other concepts are based on (somehow) similar ideas.

1. Pitman efficiency

Definition 4.1.1: Let M_o be a family of d.f. G and let T be an estimator for the estimable parameter $\vartheta = \vartheta(G)$ based on n i.i.d. observations. If $F \in M_o$ has finite Fisher information

$$I(F) = \int \left(\frac{f'(x)}{f(x)} \right)^2 dF(x) < \infty$$

then T is called *efficient* at F if

$$\text{(4.1.1)} \qquad \operatorname{Var}_F(T) = n^{-1} I(F)^{-1} .$$

A sequence $\{ T_n : n \geq 1 \}$ of estimators for $\vartheta(G)$ is called *asymptotically efficient* at F if

$$\text{(4.1.2)} \qquad \lim_{n\to\infty} n\, I(F) \operatorname{Var}_F(T_n) = 1.$$

Definition 4.1.2: Let M_o be a family of d.f. G and let $T = \{ T_n : n \geq 1 \}$ and $S = \{ S_n : n \geq 1 \}$ be two sequences of estimators of the same parameter $\vartheta = \vartheta(G)$. The *asymptotic relative efficiency of S relative to T* at F (for short: the ARE) is defined to be

$$\text{(4.1.3)} \qquad e_V(S,T) = \lim_{n\to\infty} \frac{\operatorname{Var}_F(S_n)}{\operatorname{Var}_F(T_n)}$$

provided this limit exists. If $e_V(S,T) = 1$, then S and T are said to be *asymptotically relative efficient*.

The sequence S may be called 'better' than T if the risk of S_n is smaller than that of T_n for each $n \in \mathbb{N}$ or at least if $n \longrightarrow \infty$, where the risk is defined by the quadratic loss function. If this holds asymptotically, then $e_V(S,T) \leq 1$, provided the limit exists. However, the point of view of efficiency is different from what is suggested by the previous definition. Assign to S_n and T_n real numbers a_n and b_n (say $a_n = \mathrm{Var}_F\,(S_n)$ and $b_n = \mathrm{Var}_F\,(T_n)$) and suppose that these numbers measure the goodness of the statistics. Now fix $n \in \mathbb{N}$ and define $m = m(n)$ to be that (minimal) integer for which $a_n \simeq b_m$. Then $\lim_{n\to\infty} nm^{-1}$ defines an efficiency concept provided the limit exists. We shall show first that the Pitman asymptotic relative efficiency of Definition 4.1.2 fits into such a concept.

<u>Theorem 4.1.1:</u> Let $T = \{ T_n : n \geq 1 \}$ and $S = \{ S_n : n \geq 1 \}$ be two sequences of statistics with Var T_n , Var $S_n > 0$ $(n \geq 1)$. For $n \in \mathbb{N}$ define

$$m = m(n) = \begin{cases} \min \ \{ r : \mathrm{Var}\, T_r \leq \mathrm{Var}\, S_n \} \\ \infty \qquad \text{if the above set is empty.} \end{cases}$$

If $n^{1/2}(T_n - ET_n)$ and $n^{1/2}(S_n - ES_n)$ converge to $\mathrm{N}(0,\sigma_T^2)$ and $\mathrm{N}(0,\sigma_S^2)$ respectively in the D_2-metric (cf. Definition 3.4.1), then

$$(4.1.4) \qquad e_V(S,T) = \sigma_S^2\,\sigma_T^{-2} = \lim_{n\to\infty} \frac{n}{m(n)} .$$

Especially, both limits exist.

<u>Proof:</u> The convergence in D_2 implies that

$$\lim_{n\to\infty} n\,\mathrm{Var}\,T_n = \sigma_T^2 \quad \text{and} \quad \lim_{n\to\infty} n\,\mathrm{Var}\,S_n = \sigma_S^2.$$

Let $\varepsilon > 0$ and choose $N_o \in \mathbb{N}$ such that $|\sigma_T^2 - n\,\mathrm{Var}\,T_n| < \varepsilon$ and $|\sigma_S^2 - n\,\mathrm{Var}\,S_n| < \varepsilon$ $(n \geq N_o)$. We claim that there exists an $N_1 \in \mathbb{N}$ such that $n \geq N_1$ implies $m(n) \geq N_o + 1$. Otherwise for some sequence $n_k \longrightarrow \infty$ we would have $m(n_k) = m \leq N_o$ (a constant) and therefore

$$0 < \mathrm{Var}\,T_m \leq \mathrm{Var}\,S_{n_k} \simeq n_k^{-1}\,\sigma_S^2 \longrightarrow 0,$$

a contradiction.

Thus, if $n \geq N_1$,

$$\frac{\text{Var } S_n - n^{-1}\varepsilon}{\text{Var } T_{m(n)} + n(m)^{-1}\varepsilon} \leq \frac{n^{-1}\,\sigma_S^2}{n(m)^{-1}\,\sigma_T^2}$$

and

$$\liminf_{n \longrightarrow \infty} \frac{n^{-1}\,\sigma_S^2}{n(m)^{-1}\,\sigma_T^2} \geq 1.$$

Similarly using $\text{Var } S_n < \text{Var } T_{m(n)-1}$,

$$\limsup_{n \longrightarrow \infty} \frac{n^{-1}\,\sigma_S^2}{n(m)^{-1}\,\sigma_T^2} \leq 1.$$

In the preceding theorem it would not be sufficient to assume weak convergence alone. However, using a different notion of efficiency for sequences S_n and T_n $(n \geq 1)$ of statistics yields an analoguous statement. Replace in (4.1.2) and (4.1.3) the variances by the variance of the limit distribution, i.e.

(4.1.2a) $$\sigma^2 = I(F)^{-1}$$

where $n^{1/2}(T_n-\mu_n) \longrightarrow N(0,\sigma^2)$ weakly for some sequence μ_n and $\sigma^2 > 0$, and

(4.1.3a) $$e(S,T) = \frac{\tau^2}{\sigma^2}$$

where in addition $n^{1/2}(S_n-\nu_n) \longrightarrow N(0,\tau^2)$ weakly for some sequence ν_n and $\tau^2 > 0$.

To give at least an intuitive interpretation of the meaning of (4.1.3a), σ^2 and τ^2 can be used to construct confidence intervals for T_n and S_n of certain lengths b_n and a_n at a given level. Thus we may take a_n and b_n as a criterion for the goodness of the estimators. Carrying out this programm in detail yields a statement analogous to Theorem 4.1.1.
Let us consider some examples.

Example 4.1.1: Let $\{ F(\cdot,\vartheta) : \vartheta \in \mathbb{R} \}$ be a location model, where $F = F(\cdot,0)$ has a symmetric density and finite Fisher information $I(F)$. Consider M-estimators T_n for ϑ defined by some function ψ in Definition 2.3.1, such that the assumptions in Theorem 2.3.4 are fulfilled. According to this theorem we have

$$n^{1/2}\ (T_n - t_o) \longrightarrow \mathcal{N}(0,\sigma^2)$$

weakly, where

$$\sigma^2 = (\lambda_F'(t_o))^{-2} \int \psi^2(x-t_o)\ dF(x) \quad .$$

Recall that t_o satisfies $\lambda_F(t_o) = 0$ and that

$$\lambda_F(t) = \int \psi(x-t)\ dF(x) \ .$$

By Theorem 2.3.3 the M-estimators T_n are consistent so that $t_o = 0$. Since

$$\frac{d}{dt}\lambda_F(t)\Big|_{t=0} = \frac{d}{dt}\left\{ \int \psi(x)\ f(x+t)\ dx \right\}\Big|_{t=0} =$$

$$= \int \psi(x)\ f'(x)\ dx$$

(if the interchanging or differentiation and integration is justified), it follows that

$$\sigma^2 = \left(\int \psi(x)\ f'(x)\ dx \right)^{-2} \int \psi^2(x)\ dF(x) \ .$$

If we choose $\psi = f'/f$, then (4.1.2a) holds:

$$\sigma^2 = I(F)^{-1}$$

and the sequence T_n is asymptotically efficient at F in our weaker sense.

Another estimator for ϑ is $S_n = n^{-1} \sum_{i=1}^{n} X_i$, the mean. Since $n^{1/2}\ S_n \longrightarrow \mathcal{N}(0,\tau^2)$ in the D_2-metric under F, where

$\tau^2 = \int x^2\ dF(x)$ (assuming second moment for F), the ARE in the sense of (4.1.3a) is

$$e(S,T) = I(F) \int x^2\ dF(x) \ .$$

Moreover, since lim n Var $T_n \geqq I(F)^{-1}$,

$$e_V(S,T) \leqq I(F) \int x^2\ dF(x)$$

provided the limit $e_V(S,T)$ exists.

For example, if F is normal $\mathcal{N}(0,a^2)$, we obtain $I(F) = a^{-2}$ and hence $e(S,T) = 1$. A simple example for which $e(S,T) > 1$ is given by the exponential distribution.

Example 4.1.2: Let $T_n = \sum_{1 \le i \le n} a(R_i^+) \operatorname{sign} X_i$ be a sequence of signed rank statistics with increasing scores a(i) defined by $a(i) = h(i/(n+1))$ where $h \in \mathcal{H}$ is absolutely continuous. Denote by $\hat{\vartheta}_n$ the corresponding R-estimator defined in Definition 3.5.2 (3.5.25). If $\{F(\cdot,\vartheta) : \vartheta \in \mathbb{R}\}$ is a location model with symmetric, absolutely continuous d.f. $F = F(\cdot,0)$, Theorem 3.5.6 asserts that

$$n^{1/2}\, \hat{\vartheta}_n \longrightarrow \mathcal{N}(0, A^{-2})$$

weakly where

$$1/A = \sqrt{2}\, \sigma \left(\int g'(F(t))\, f^2(t)\, dt \right)^{-1},$$

$$\sigma^2 = \int_{1/2}^{1} h^2(2t-1)\, dt$$

and

$$g(t) = \begin{cases} h(2t-1) & \text{if } t \ge 1/2 \\ -h(1-2t) & \text{if } t < 1/2. \end{cases}$$

(Here we assume that the hypothesis of Theorem 3.5.6 is fulfilled.) According to Example 3.5.6 $\sigma^2 = I(F)^{-1}$ if

$$h(t) = -(f'/f) \circ F^{-1}(1/2 + t/2)$$

and by Example 3.5.8

$$A^2 = I(F) \quad .$$

It follows that $\hat{\vartheta}_n$ is asymptotically efficient in the sense of (4.1.2a) . If $S_n = n^{-1} \sum_{1 \le i \le n} X_i$ denotes the mean as before, then

$$e(S,\hat{\vartheta}) = I(F) \int x^2\, dF(x).$$

Also we have $e(T,\hat{\vartheta}) = 1$ for the M-estimator of Example 4.1.1.

Example 4.1.3: Let now $L_n = \sum_{1 \le i \le n} c_i X_{(i)}$ be a sequence of L-estimators (cf. Definition 3.6.1) where $c_i = g(i/(n+1))$ is defined by some fixed score function $g \in \mathcal{H}$. If the assumptions of Theorem 3.6.2 are satisfied then

$$n^{1/2}(L_n - \int F^{-1}(t)\, g(t)\, dt)$$

is asymptotically normal $\mathcal{N}(0,\sigma^2)$ in the D_2-metric, where

$$\sigma^2 = 2 \iint_{s<t} s(1-t)\, g(t)\, g(s)\, dF^{-1}(s)\, dF^{-1}(t) \; .$$

Assume furthermore that F is symmetric and has finite Fisher information I(F). Setting

$$g(x) = -\frac{1}{I(F)} \frac{f''f-(f')^2}{f^2} \circ F^{-1}(x),$$

$$\int_{-\infty}^{\infty} F^{-1}(t,\vartheta) \; g(t) \; dt =$$

$$= -\frac{1}{I(F)} \int_{-\infty}^{\infty} t \left(\frac{f''f-(f')^2}{f^2}\right)(t-\vartheta) \; f(t-\vartheta) \; dt = \vartheta$$

by symmetry, where $F(x,\vartheta) = F(x-\vartheta)$ $(x \in \mathbb{R})$. Also

$$\sigma^2 = \frac{2}{I(F)^2} \iint_{s<t} F(s)(1-F(t)) \frac{f''f-(f')^2}{f^2}(s) \frac{f''f-(f')^2}{f^2}(t) \; ds \; dt =$$

$$= \frac{n}{I(F)^2} \iint E(\hat{H}_n(s)-F(s))(\hat{H}_n(t)-F(t)) \frac{f''f-(f')^2}{f^2}(s) \frac{f''f-(f')^2}{f^2}(t) \; ds \; dt =$$

$$= \frac{n}{I(F)^2} E\left(\int \frac{f'(s)}{f(s)} \; d(\hat{H}_n-F)(s)\right)^2 =$$

$$= I(F)^{-2} \; \mathrm{Var}_F \; (f'/f) = I(F)^{-1}.$$

It follows that L_n is asymptotically efficient (in the sense of (4.1.2)!). If S_n denotes the mean of $X_1,\ldots,X_n$ then

$$e_V(S,L) = I(F) \int x^2 \; dF(x)$$

and this equals 1 if F is normal.
Let F be defined by its density

$$f(x) = \begin{cases} \alpha/2 \; \exp(-\alpha x) & \text{if } x \geq 0 \\ \alpha/2 \; \exp(\alpha x) & \text{if } x < 0. \end{cases}$$

Then $I(F) = \alpha^2$ and $\int x^2 \; dF(x) = 2\alpha^{-2}$. In this case $e_V(S,L) = 2$.
Also note that $e(T,L) = 1$ and $e(\hat{\vartheta},L) = 1$ where T and $\hat{\vartheta}$ were defined in Examples 4.1.1 and 4.1.2.
We have seen in the last three examples that M-, R- and L-estimators provide asymptotically efficient estimation procedures. We may summarize in the following way:

- an M-estimator is the solution of $\int \psi(x-T(F))\, dF(x) = 0$ and is asymptotically efficient for $\psi = -(f'/f)$.
- an R-estimator is the solution of

$$\int g(1/2\ (F(x)+1-F(2T(F)-x)))\, dF(x) = 0$$

  and it is asymptotically efficient if $g = -(f'/f)\circ F^{-1}$.
- an L-estimator is defined by $T(F) = \int g(t)\, F^{-1}(t)\, dt$ and it is asymptotically efficient if

$$g(t) = I(F)^{-1} \frac{d}{ds}\left\{-\frac{f'(s)}{f(s)}\right\}\Big|_{s=F^{-1}(t)} .$$

Their ARE's in the sense of (4.1.2a) are all 1. But the L-estimators are asymptotically efficient in a stronger sense. Under slightly more restrictive assumptions the same result holds for R-estimators. Obviously it is sufficient to prove (3.5.24) in Theorem 3.5.5 in the D_2-metric. This is contained in the following result for R-estimators.

<u>Theorem 4.1.2:</u> Let the assumptions of Theorem 3.5.6 be satisfied. Assume in addition that $h - h(1/2) \in H_\delta$ for some $\delta > 0$ (cf. (3.6.4)). Then $g \in H_\delta$ and

$$E\, |T_N(g)|^{2+\delta} \leq K\, \|g\|_\delta^{2+\delta} \tag{4.1.5}$$

for every $N \in \mathbb{N}$ where K is some constant.
Moreover, (3.5.26) holds in the D_2-metric, if F satisfies $1 - F(x) \leq \text{const.}\ x^{-a}$ for some $a > 0$ and all sufficiently large x.

<u>Proof:</u> Recall that $g(t) = h(2t-1)$ $(t \geq 1/2)$ and $g(t) = -h(1-2t)$ $(t < 1/2)$. Since $h(0) = 0$ we have $g \in H_\delta$.
Let us first show (4.1.5). We will follow the proof of the Approximation Theorem 3.4.2. Especially, let $\varphi(z,t)$ denote the function defined in its proof. Let $\psi(z) = \int \varphi(z,t)\, d\hat{F}_N(t)$ and $u(z) = (z(1-z))^{\delta/(2+\delta)^2}$. Writing $p = 1 + \delta/2$ and $q = 1 + 2\delta^{-1}$ we first obtain

$$N^p\, E\, \Big| \int g(\tfrac{N}{N+1}\hat{H}_N(t)) - g(\tfrac{N}{N+1}H_N(t))\, d\hat{F}_N(t)\, \Big|^{2p} =$$

$$= N^p\, E\, \Big| \int_0^{N/N+1} \int \varphi(z,t)\, d\hat{F}_N(t)\, dg(z) \Big|^{2p} \leq$$

$$\leqq N^p \, E \left| \int_0^{N/N+1} u(z)^q \, dg(z) \right|^{2p/q} \left(\int_0^{N/N+1} \left| \frac{\psi(z)}{u(z)} \right|^p dg(z) \right)^2 =$$

$$= N^p \left(\int_0^{N/N+1} (z(1-z))^{1/2 - \eta} \, dg(z) \right)^\delta$$

$$\int_0^{N/N+1} \int_0^{N/N+1} E \frac{|\psi(z) \; \psi(y)|^p}{u(z)^p u(y)^p} \, dg(z) \, dg(y) <<$$

$$<< N^p \|g\|_\delta^\delta \left(\int_0^{N/N+1} u(z)^{-p} (E|\psi(z)|^{2p})^{1/2} \, dg(z) \right)^2 <<$$

$$<< N^p \|g\|_\delta^\delta \left(\int_0^{N/N+1} (z(1-z))^{-\eta} \left\{ \min \{N^{-2p}, z^{2p}\} + N^{-p}(z(1-z)) \right\}^{1/2} dg(z) \right)^2 <<$$

$$<< \|g\|_\delta^{2+\delta}.$$

Here $a << b$ means that $a \leqq cb$ for some constant c and η is as in (3.6.4).

Next observe that

$$N^{1+\delta/2} E \left| \int g(\tfrac{N}{N+1} H_N(t)) \, d(\hat{F}_N - F_N)(t) \right|^{2+\delta} <<$$

$$<< \int |g(t)|^{2+\delta} \, dt \; << \|g\|_\delta^{2+\delta}$$

by Lemma 3.6.1, since for i.i.d. variables $Y_1, \dots, Y_n$ with $EY_1 = 0$ and $E|Y_1|^{2+\delta} < \infty$ one has

$$E \left| \sum_{i=1}^n Y_i \right|^{2+\delta} \leqq \text{const. } n^{1+\delta/2} \, E|Y_1|^{2+\delta},$$

where the constant does not depend on Y_1.

Finally,

$$N^{1+\delta/2} \left| \int_{-\infty}^{\infty} (g(H_N(t)) - g(\tfrac{N}{N+1} H_N(t)) \, dF_N(t) \right|^{2+\delta} <<$$

$$<< \|g\|^{2+\delta} << \|g\|_\delta^{2+\delta}$$

has been shown in the proof of Theorem 3.4.2.

Putting everything together proves (4.1.5).

From Theorem 3.5.6 we know that $n^{1/2} \hat{\vartheta}_n \longrightarrow N(0, A^{-2})$ weakly

where $A = 2^{-1/2} \sigma^{-1} \int g'(F(t)) \, f^2(t) \, dt$. This weak congence is also true in the D_2-metric if we can show that

$$(4.1.6) \qquad \lim_{n \to \infty} n \, E \, \hat{\vartheta}_n^2 = A^{-2}.$$

Indeed, if (4.1.6) holds, then $\{n\hat{\vartheta}_n^2 : n \geq 1\}$ is uniformly integrable. The asymptotic normality of $n^{1/2}\hat{\vartheta}_n$ also ensures the existence of versions of $n^{1/2}\hat{\vartheta}_n$ and of an $N(0,A^{-2})$-distributed random variable Y, defined on one probability space, such that $n^{1/2}\hat{\vartheta}_n - Y \longrightarrow 0$ in probability. This convergence and the uniform integrability imply that the convergence also holds in L_2-norm, equivalently

$$D_2(\mathcal{L}(n^{1/2}\hat{\vartheta}_n), N(0,A^{-2})) \longrightarrow 0.$$

Let Λ_n denote the d.f. of $n^{1/2}\hat{\vartheta}_n$, which is continuous according to (3.5.23). Since Λ_n is symmetric and since $E\,\hat{\vartheta}_n^2 < \infty$,

$$n\,E\,\hat{\vartheta}_n^2 = \int_{-\infty}^{\infty} x^2\, d\Lambda_n(x) = 4\int_0^{\infty} x(1-\Lambda_n(x))\,dx .$$

Because of the weak convergence of Λ_n to $N(0,A^{-2})$ the functions $x \longrightarrow x(1-\Lambda_n(x))$ converge pointwise to $x(1-\Lambda(x))$ where Λ denotes the d.f. of $N(0,A^{-2})$. Since

$$A^{-2} = 4\int_0^{\infty} x(1-\Lambda(x))\,dx ,$$

(4.1.6) follows, if the functions $x(1-\Lambda_n(x))$ are bounded by some integrable function.

In order to prove this (for sufficiently large n) we shall define three constants c, $K_1 \leq K_2$ and prove

$$x(1-\Lambda_n(x)) \leq c\,x^{-1-\delta} \qquad (n \geq n_o) \tag{4.1.7}$$

separately on each of the intervals $[1,K_1n^{1/2}]$, $(K_1n^{1/2},K_2n^{1/2})$ and $[K_2n^{1/2},\infty)$.

Let us consider the first interval, where we define K_1 at the same time. Recall from Theorem 3.5.6 that for

$$\mu_n(u) = 2^{-1}\int g(1/2\,(1+F_u(t)-F_u(-t)))\,dF_u(t)$$

we have

$$\lim_{u\to 0} u^{-1}\mu_n(u) = \int g'(F(t))\,f^2(t)\,dt > 0.$$

Hence there exist constants $K_1 > 0$ and $c_o > 0$ such that $|u| \leq K_1$ implies $u^{-1}\mu_n(u) \geq c_o > 0$.

Recall also from the proof of Theorem 3.5.5 that

$$\{n^{1/2}\hat{\vartheta}_n \geq x\} \subset \{y \in \mathbb{R}^n: T_n((y_1 - xn^{-1/2}, \dots, y_n - xn^{-1/2})) \geq 0\}.$$

It follows from the Markov inequality and (4.1.5) that

$$1 - \Lambda_n(x) \leq F_{-xn^{-1/2}}(\{T_n - \mu_n(-xn^{-1/2}) \geq -\mu_n(-xn^{-1/2})\}) \leq$$

$$\leq \text{const.}\ |\mu_n(-xn^{-1/2})|^{-2-\delta}\ n^{-1-\delta/2} \|g\|_\delta^{2+\delta}.$$

If $x \leq K_1 n^{1/2}$, one obtains

$$x(1 - \Lambda_n(x)) \leq \text{const.}\ \|g\|_\delta^{2+\delta}\ c_0^{-2-\delta}\ x^{-1-\delta} \leq c\ x^{-1-\delta},$$

proving (4.1.7) for all n and $x \leq K_1 n^{1/2}$.

Next consider the third interval, defining K_2 at the same time. Choose $m(n) \in \mathbb{N}$ minimal in order to satisfy

$$-\sum_{i=1}^{m(n)} h(i/n+1) + \sum_{i=m(n)+1}^{n} h(i/n+1) < 0.$$

Since $h(0) = 0$ and since h is increasing and integrable, $\lim m(n)/n = \gamma$ exists and $m(n)/n \geq 1/2$ for sufficiently large n. (In fact γ is determined by $\int_{(0,\gamma)} h(t)dt = \int_{(\gamma,1)} h(t)dt$.)

Fix $y = (y_1, \dots, y_n) \in \mathbb{R}^n$ and let $y^* = y_{(m(n))}$. It follows that $T_n((y_1 - y^*, \dots, y_n - y^*)) < 0$, i.e. $\hat{\vartheta}_n(y) \leq y_{(m(n))}$. This implies that

$$1 - \Lambda_n(xn^{1/2}) = F^n(\hat{\vartheta}_n > x) \leq P(X_{(m(n))} > x) =$$

$$= \sum_{k=n-m(n)+1}^{n} \binom{n}{k} (1 - F(x))^k F(x)^{n-k} =$$

$$= \binom{n}{n-m(n)+1} (n-m(n)+1) \int_0^{1-F(x)} t^{n-m(n)} (1-t)^{m(n)-1}\, dt \leq$$

$$\leq \binom{n}{n-m(n)+1} (1 - F(x))^{n-m(n)+1} \leq n^{-1/2} \left\{ \frac{1-F(x)}{C(\gamma)} \right\}^{n(1-\gamma)},$$

using Stirling's formula and $(m(n)-1)n^{-1} \geq 1/2$ if n is large enough. Here $C(\gamma)$ denotes some constant depending on γ.

Since by assumption $1-F(x) \leq \text{const.}\ x^{-a}$ for all sufficiently large x, there exists a constant $K_2 \geq K_1$ such that

$$1 - \Lambda_n(xn^{1/2}) \leq n^{-1/2}\ x^{-1/2\ an(1-\gamma)} \qquad (x \geq K_2).$$

If $x \geq K_2 n^{1/2}$ and if n is large enough, $an(1-\gamma) \geq 2+\delta$ and $\frac{1+\delta}{an(1-\gamma)} \log n \leq 1/2 \log K_2$, then

$$n^{-1/2}\ (xn^{-1/2})^{-1/2\ an(1-\gamma)} \leq x^{-2-\delta}$$

and therefore

$$x(1 - \Lambda_n(x)) \leq x^{-1-\delta} .$$

Finally in the interval $K_1 n^{1/2} < x < K_2 n^{1/2}$ use $\frac{K_1}{K_2} \leq 1$ and the first case to obtain

$$x(1 - \Lambda_n(x)) \leq x(1- \Lambda_n(x\, K_1/K_2)) \leq$$

$$\leq \text{const.}\ \|g\|_\delta^{2+\delta}\ c_o^{-2-\delta}\ (K_2/K_1)^{2+\delta}\ x^{-1-\delta} \leq c\ x^{-1-\delta}.$$

This completes the proof of (4.1.7) and of the theorem.

In the remaining part of this section the Pitman efficiency for test statistics will be considered, but only in a rather special (though the most important) case.

Definition 4.1.3: Let $\{F_\vartheta : \vartheta \in \mathbb{R}\}$ be a family of d.f. and let $\vartheta_o \in \mathbb{R}$ be fixed. A sequence $T = \{ T_n : n \in \mathbb{N} \}$ of statistics $T_n : \mathbb{R}^n \longrightarrow \mathbb{R}$ is said to satisfy condition $A(L,\delta)$ where $L, \delta > 0$, if the following holds:

There exist functions $\mu_n : \mathbb{R} \longrightarrow \mathbb{R}$ and $\sigma_n : \mathbb{R} \longrightarrow \mathbb{R}$ $(n = 1,2,\ldots)$ such that conditions (4.1.8) - (4.1.11) are satisfied.

(4.1.8) The functions σ_n are uniformly continuous at ϑ_o and

$$0 < \sigma_n^2(\vartheta) < \infty \qquad (\vartheta_o \leq \vartheta,\ n \geq 1).$$

(4.1.9) $\sigma_n^{-1}(\vartheta_o)\ (T_n - \mu_n(\vartheta_o)) \longrightarrow \mathcal{N}(0,1)$ weakly under F_{ϑ_o}.

(4.1.10) For each $n \geq 1$, the function μ_n is differentiable on $[\vartheta_o,\infty)$ with derivative μ_n' satisfying

(a) $\mu_n'(\vartheta_o) > 0$.

(b) The functions μ_n' $(n \in \mathbb{N})$ are uniformly continuous at ϑ_o.

(c) $\lim\limits_{n\to\infty} \mu_n'(\vartheta_o)\ n^{-\delta}\ \sigma_n^{-1}(\vartheta_o) = L > 0.$

(4.1.11) $\sigma_n^{-1}(\vartheta_n)(T_n - \mu_n(\vartheta_n)) \longrightarrow \mathcal{N}(0,1)$ weakly under $F_{\vartheta_n}^n$

for every sequence $\vartheta_n = \vartheta_o + \tau\, n^{-\delta} + o(n^{-\delta})$ $(\tau > 0)$.

Note that in the previous definition $\delta > 0$ is uniquely determined by (4.1.10)(c). Also we would like to remark that in (4.1.11) the sample sizes and the distributions F_{ϑ_n} (the alternatives) vary.

Lemma 4.1.1: Let $T = \{T_n : n \in \mathbb{N}\}$ satisfy condition $A(L,\delta)$, and let $\{\alpha_n : n \in \mathbb{N}\}$ be a sequence of real numbers $0 < \alpha_n < 1$ satisfying $\lim \alpha_n = \alpha$ for some $0 < \alpha < 1$. Using T_n and α_n define k_n by the level α_n test φ_n, where $0 \leq \gamma_n \leq 1$,

$$(4.1.12) \qquad \begin{cases} \varphi_n = \gamma_n 1_{\{T_n = k_n\}} + 1_{\{T_n > k_n\}} \\ E_{\vartheta_o}(\varphi_n) = \alpha_n \end{cases} \qquad (n \geqq 1).$$

Let $\vartheta_n \searrow \vartheta_o$ be a sequence of alternatives as in (4.1.11).

Setting $h_n = \frac{k_n - \mu_n(\vartheta_n)}{\sigma_n(\vartheta_n)}$ and $g_n = \frac{k_n - \mu_n(\vartheta_o)}{\sigma_n(\vartheta_o)}$

we have

$$(4.1.13) \qquad \lim_{n\to\infty} g_n = g = \Phi^{-1}(1-\alpha)$$

and

$$(4.1.14) \qquad \lim_{n\to\infty} h_n = h = g - \tau L$$

where Φ denotes the d.f. of the standard normal distribution. Moreover, the power $\beta_n(\vartheta_n) = E_{\vartheta_n}(\varphi_n)$ of φ_n at ϑ_n satisfies

$$(4.1.15) \qquad \lim_{n\to\infty} \beta_n(\vartheta_n) = \beta = 1 - \Phi(h).$$

(Note that $h < g$ and thus $\beta > \alpha$.)

Proof: By (4.1.9)

$$\alpha_n = E_{\vartheta_o}\varphi_n = \gamma_n F^n_{\vartheta_o}\left\{ \frac{T_n - \mu_n(\vartheta_o)}{\sigma_n(\vartheta_o)} = g_n \right\} +$$

$$+ F^n_{\vartheta_o}\left\{ \frac{T_n - \mu_n(\vartheta_o)}{\sigma_n(\vartheta_o)} > g_n \right\} \longrightarrow 1 - \lim_{n\to\infty} \Phi(g_n)$$

$$= 1 - \Phi(g)$$

and (4.1.13) follows.

Now by (4.1.8) and (4.1.10) (b),(c) and by Taylor's theorem

$$h_n = g_n \sigma_n(\vartheta_o)\sigma_n^{-1}(\vartheta_n) + \sigma_n^{-1}(\vartheta_n)(\mu_n(\vartheta_o) - \mu_n(\vartheta_n)) =$$

$$= \sigma_n(\vartheta_o)\sigma_n^{-1}(\vartheta_n)(g_n - \tau n^{-\delta}\mu_n'(\xi(\vartheta_n))\sigma_n^{-1}(\vartheta_o)) + o(1)$$

$$\longrightarrow g - L\tau ,$$

proving (4.1.14).

Finally, by (4.1.11)

$$\beta_n(\vartheta_n) = \gamma_n F^n_{\vartheta_n}\left\{ \frac{T_n - \mu_n(\vartheta_n)}{\sigma_n(\vartheta_n)} = h_n \right\} +$$

$$+ F^n_{\vartheta_n} \left\{ \frac{T_n - \mu_n(\vartheta_n)}{\sigma_n(\vartheta_n)} > h_n \right\} \longrightarrow 1 - \Phi(h) = \beta .$$

Let $T = \{T_n : n \geq 1\}$ and $S = \{S_n : n \geq 1\}$ be two sequences satisfying the condition $A(L,\delta)$ and $A(\tilde{L},\tilde{\delta})$ respectively. Let $\{ F_\vartheta : \vartheta \in \mathbb{R} \}$ and $\vartheta_o \in \mathbb{R}$ be as in Definition 4.1.3. Then we set

(4.1.16) $$\tilde{e}(T,S) = \tilde{L}/L \lim_{n\to\infty} n^{\tilde{\delta}-\delta} \in [0,\infty].$$

We observe the following

<u>Theorem 4.1.3:</u> Let $\{ F_\vartheta : \vartheta \in \mathbb{R} \}$ and $\vartheta_o \in \mathbb{R}$ be as before. Assume that $T = \{ T_n : n \geq 1 \}$ and $S = \{ S_n : n \geq 1 \}$ satisfy the condition $A(L,\delta)$ and $A(\tilde{L},\tilde{\delta})$ respectively where $L,\tilde{L},\delta,\tilde{\delta} > 0$. Let $0 < \alpha < 1$ be given and denote by φ_n (resp. $\tilde{\varphi}_n$) a sequence of tests obtained by using T_n (resp. S_n) according to (4.1.12) such that $\lim E_{\vartheta_o}(\varphi_n) = \lim E_{\vartheta_o}(\tilde{\varphi}_n) = \alpha$.

(1) If $\alpha < \beta < 1$, then there exists a sequence $\vartheta_n = \vartheta_o + \tau n^{-\delta} + o(n^{-\delta})$ for some $\tau > 0$ such that

$$\lim_{n\to\infty} E_{\vartheta_n}(\varphi_n) = \beta .$$

(2) If β and ϑ_n are as in (1), then there exists a sequence $m(n) \longrightarrow \infty$ such that $\vartheta_n = \vartheta_o + \tilde{\tau}\, m(n)^{-\tilde{\delta}} + o(m(n)^{-\tilde{\delta}})$ and

$$\lim_{n\to\infty} E_{\vartheta_n}(\tilde{\varphi}_{m(n)}) = \beta .$$

(3) Assume that the asymptotic normality condition in (4.1.11) holds for every sequence $\vartheta_n \searrow \vartheta_o$ and for T as well as for S. Let $0 < \alpha < \beta < 1$.
If $\vartheta_n \searrow \vartheta_o$ satisfies $\lim E_{\vartheta_n}(\varphi_n) = \beta$ and if $\{ m(n) : n \geq 1 \}$ is a sequence of integers $m(n)$ satisfying $m(n) \longrightarrow \infty$ and $\lim E_{\vartheta_n}(\tilde{\varphi}_{m(n)}) = \beta$, then

(4.1.17) $$\tilde{e}(T,S) = \lim_{n\to\infty} \left(\frac{n}{m(n)} \right)^{\tilde{\delta}} .$$

<u>Proof:</u> (1) follows immediately from (4.1.14) and (4.1.15) applied to T, choosing $\tau = L^{-1} (\Phi^{-1}(1-\alpha) - \Phi^{-1}(1-\beta))$.

(2) Define
$$m(n) = \left[\left(\frac{L}{\tilde{L}} \right)^{1/\tilde{\delta}} n^{\delta/\tilde{\delta}} \right] .$$

Then $\vartheta_n = \vartheta_o + \tau(L/\tilde{L})\, m(n)^{-\tilde{\delta}} + o(m(n)^{-\tilde{\delta}})$ and by (4.1.11)

$$\lim_{n\to\infty} E_{\vartheta_n}(\tilde{\varphi}_{m(n)}) = \lim_{n\to\infty} F^n_{\vartheta_n}\left\{ \frac{S_m-\tilde{\mu}_m(\vartheta_n)}{\tilde{\sigma}_m(\vartheta_n)} > \frac{\tilde{k}_m-\tilde{\mu}_m(\vartheta_n)}{\tilde{\sigma}_m(\vartheta_n)} \right\} = \beta$$

where $m = m(n)$, since by (4.1.8), (4.1.10) and (4.1.13)

$$\lim_{n\to\infty} \frac{\tilde{k}_m-\tilde{\mu}_m(\vartheta_n)}{\tilde{\sigma}_m(\vartheta_n)} = \lim_{n\to\infty} \frac{\tilde{k}_m-\tilde{\mu}_m(\vartheta_o)}{\tilde{\sigma}_m(\vartheta_o)} - \lim_{n\to\infty} \tilde{\mu}'_m(\vartheta_o)\,\tilde{\sigma}_m^{-1}(\vartheta_o)(\vartheta_n-\vartheta_o)$$

$$= g - \tau\, L.$$

(3) Let β, $\vartheta_n \searrow \vartheta_o$ and $m(n)$ be given as in (3), and denote by $k_n, h_n, g_n, \tilde{k}_n, \tilde{h}_n, \tilde{g}_n$ the numbers defined in Lemma 4.1.1 for T and S respectively. We repeat some of the previous arguments. By asymptotic normality,

$$1 - \Phi(\alpha) = \lim E_{\vartheta_o}(\varphi_n) = \lim F^n_{\vartheta_o}\left\{ \frac{T_n-\mu_n(\vartheta_o)}{\sigma_n(\vartheta_o)} > g_n \right\}$$

and therefore $g = \lim g_n$ exists. Similarly

$$g = \lim \tilde{g}_n = \Phi^{-1}(1-\alpha)$$

exists.

Moreover (4.1.8), (4.1.10) and

$$\beta = \lim E_{\vartheta_n}(\varphi_n) = \lim F^n_{\vartheta_n}\left\{ \frac{T_n-\mu_n(\vartheta_n)}{\sigma_n(\vartheta_n)} > h_n \right\}$$

imply that

$$\Phi^{-1}(1-\beta) = \lim h_n = g - \lim \mu'_n(\vartheta_o)\sigma_n^{-1}(\vartheta_o)\ (\vartheta_n-\vartheta_o)$$

exists.

Writing $m = m(n)$ we see that

$$\beta = \lim E_{\vartheta_n}(\tilde{\varphi}_m) = \lim F^n_{\vartheta_n}\left\{ \frac{S_m-\tilde{\mu}_m(\vartheta_n)}{\tilde{\sigma}_m(\vartheta_n)} > \frac{\tilde{k}_m-\tilde{\mu}_m(\vartheta_n)}{\tilde{\sigma}_m(\vartheta_n)} \right\}$$

implies the existence of the limit

$$\lim \frac{\tilde{k}_m-\tilde{\mu}_m(\vartheta_n)}{\tilde{\sigma}_m(\vartheta_n)} = g - \lim \tilde{\mu}'_m(\vartheta_o)\ \tilde{\sigma}_m^{-1}(\vartheta_o)\ (\vartheta_n-\vartheta_o) =$$

$$= \Phi^{-1}(1-\beta)$$

by (4.1.8) and (4.1.10). Using (4.1.10)(c) it follows that

$$\tilde{L} \lim m^{\tilde{\delta}}\ (\vartheta_n-\vartheta_o) = L \lim n^{\delta}\ (\vartheta_n-\vartheta_o)$$

and consequently

$$(L/\tilde{L}) \lim n^{\delta} m^{-\tilde{\delta}} = 1$$

or

$$\lim (n/m)^{\tilde{\delta}} = \frac{\tilde{L}}{L} \lim (n/m)^{\tilde{\delta}} m^{\tilde{\delta}} n^{-\delta} = \tilde{e}(T,S).$$

Theorem 4.1.3 fits into the discussion after Definition 4.1.4 on the general concept of efficiency and leads to the definition of Pitman efficiency for test statistics.

Definition 4.1.4: Let $\{ F_\vartheta : \vartheta \in \mathbb{R} \}$ be a family of d.f. and let $\vartheta_o \in \mathbb{R}$. For two sequences $T = \{ T_n : n \geq 1 \}$ and $S = \{ S_n : n \geq 1 \}$ of statistics $T_n, S_n : \mathbb{R}^n \longrightarrow \mathbb{R}$ define tests φ_n (resp. $\tilde{\varphi}_n$) by (4.1.12) so that

$$\lim_{n\to\infty} E_{\vartheta_o}(\varphi_n) = \lim_{n\to\infty} E_{\vartheta_o}(\tilde{\varphi}_n) = \alpha \in (0,1).$$

For $\alpha < \beta < 1$ let $\vartheta_n \searrow \vartheta_o$ be a sequence of alternatives such that

$$\lim_{n\to\infty} E_{\vartheta_n}(\varphi_n) = \beta.$$

Consider a sequence $m(n) \longrightarrow \infty$ (as $n \longrightarrow \infty$) satisfying

$$\lim_{n\to\infty} E_{\vartheta_n}(\tilde{\varphi}_{m(n)}) = \beta.$$

If $\lim\limits_{n\to\infty} \frac{n}{m(n)}$ exists and if this limit is independent of all choices made (i.e. of α, β, $\vartheta_n \searrow \vartheta_o$ and $m(n)$), then this common limit is called the *asymptotic relative Pitman efficiency of T relative to S at* ϑ_o and denoted by $e_P(T,S)$.

Theorem 4.1.3 states that under condition $A(\cdot,\cdot)$ the relative asymptotic Pitman efficiency of two sequences T and S of statistics exists, provided (4.1.11) also holds for T and S and for all alternatives $\vartheta_n \searrow \vartheta_o$. Furthermore, $e_P(T,S)$ can be computed from (4.1.17). We give an example for this application.

Example 4.1.4: Consider the family $F_\vartheta = \mathcal{N}(\vartheta,\sigma^2)$ of normal distributions ($\vartheta \in \mathbb{R}$), where σ^2 is fixed but unknown. Let $X_1,\dots,X_n,Y_1,\dots,Y_m$ be independent random variables with distributions $\mathcal{L}(X_i) = F_{\vartheta_1}$ and $\mathcal{L}(Y_j) = F_{\vartheta_2}$. For testing $\vartheta_2 \leq \vartheta_1$ against $\vartheta_2 > \vartheta_1$ the best parametric test is the t-test

$$- T_N = (\overline{X} - \overline{Y}) \left\{\frac{N}{nm(N-2)}\right\}^{-\frac{1}{2}} \left(\sum_{i=1}^{n} (X_i - \overline{X})^2 + \sum_{j=1}^{m} (Y_j - \overline{Y})^2 \right)^{-1/2}$$

where $N = n + m$. If F_ϑ is not normal or if the normality can not be guaranteed, one could use the van der Waerden test

$$- S_N = \int_{-\infty}^{\infty} \Phi^{-1}\left(\frac{N}{N+1} \hat{H}_N(t)\right) d\hat{F}_N(t)$$

as we saw in Example 3.2.1, or the Wilcoxon two sample test

$$- W_N = \int_{-\infty}^{\infty} \frac{N}{N+1} \hat{H}_N(t) \, d\hat{F}_N(t)$$

can be applied (cf. Examples 3.2.1 and 1.4.1).
A test for $\vartheta_2 - \vartheta_1 \leqq 0$ against $\vartheta_2 - \vartheta_1 > 0$ rejects the hypothesis if $-T_N \leqq c_T$, $-S_N \leqq c_S$ or $-W_N \leqq c_W$ for certain constants c_T, c_S, c_W depending on the level. Thus, in order to be consistent with the previous definitions, especially with (4.1.12), the asymptotic relative efficiencies should be computed using T_N, S_N and W_N.

(a) T_N satisfies the condition $A(L_T, 1/2)$ with $\mu_N(\vartheta) = \vartheta$, $\mu_N'(0) = 1$ and

$$\lim N \, \sigma_N^2(0) = \lim \frac{N^3}{nm(N-2)} N^{-1} \left(\sum_{i=1}^{n} E(X_i - \overline{X})^2 + \sum_{j=1}^{m} E(Y_j - \overline{Y})^2 \right)$$

$$= (\lambda(1-\lambda))^{-1} \sigma^2$$

as $nN^{-1} \longrightarrow \lambda$ and $n, m \longrightarrow \infty$.
It follows that

$$L_T = \lim \frac{\mu_N'(0)}{N^{1/2} \sigma_N(0)} = \sigma^{-1} (\lambda(1-\lambda))^{1/2} .$$

(b) S_N satisfies the condition $A(L_S, 1/2)$ (use Theorem 3.4.3 and the following remark together with (3.4.9)) with

$$\mu_N(\vartheta) = - \int_{-\infty}^{\infty} \Phi^{-1}(H_N(t)) \, dF_N(t) =$$

$$= - \frac{n}{N} \int_{-\infty}^{\infty} \Phi^{-1}\left(\frac{n}{N} F_{\vartheta_1}(t) + \frac{m}{N} F_{\vartheta_2}(t)\right) dF_{\vartheta_1}(t)$$

$(\vartheta = \vartheta_2 - \vartheta_1)$ and

$$\sigma_N^2(0) = \frac{2}{N}(1-\lambda)\lambda \iint_{s<t} F_{\vartheta_1}(s)(1-F_{\vartheta_1}(t))\, d\Phi^{-1}\circ F_{\vartheta_1}(s)\, d\Phi^{-1}\circ F_{\vartheta_1}(t)$$

$$= \frac{2}{N}(1-\lambda)\lambda \iint_{s<t} \Phi(s)(1-\Phi(t))\, ds\, dt =$$

$$= \frac{2}{N}(1-\lambda)\lambda \iint_{s<t} \frac{1}{2\pi} \int_{-\infty}^{s} e^{-u^2/2} du \int_{t}^{\infty} e^{-v^2/2} dv\, dsdt =$$

$$= \frac{2}{N}(1-\lambda)\lambda\frac{1}{2\pi} \iiiint e^{-(u^2+v^2)/2}\, 1_{\{u\leq s<t\leq v\}}\, dsdtdudv$$

$$= \frac{1}{N}(1-\lambda)\lambda\frac{1}{2\pi} \iint_{u<v} (u-v)^2\, e^{-(u^2+v^2)/2}\, du\, dv =$$

$$= \frac{\lambda(1-\lambda)}{N} .$$

Let $F = F_o$. In order to compute $\mu_N'(0)$ write $\mu_N(\vartheta)$ in the form

$$\mu_N(\vartheta) = -\frac{n}{N} \int_{-\infty}^{\infty} \Phi^{-1}\left(\frac{n}{N} F(t) + \frac{m}{N} F(t-\vartheta)\right) dF(t).$$

It follows that

$$\mu_N'(\vartheta) = \frac{nm}{N^2} \int_{-\infty}^{\infty} \left\{\frac{1}{(2\pi)^{1/2}} \exp - \frac{(\Phi^{-1}(\frac{n}{N}F(t)+\frac{m}{N}F(t-\vartheta)))^2}{2}\right\}^{-1}$$

$$(2\pi\sigma^2)^{-1} \exp(-(t-\vartheta)^2/2\sigma^2) \exp(-t^2/2\sigma^2)\, dt.$$

Using $\Phi^{-1}(F(t)) = t/\sigma$ the last equality can be evaluated for $\vartheta = 0$, that is

$$\mu_N'(0) = \frac{nm}{N^2} \int_{-\infty}^{\infty} \left((2\pi)^{-1/2} \exp(-t^2/2\sigma^2)\right)^{-1}$$

$$\left[(2\pi\sigma^2)^{-1/2} \exp(-t^2/2\sigma^2)\right]^2 dt =$$

$$= \frac{nm}{\sigma N^2} .$$

It follows now that

$$L_S = \lim (\lambda(1-\lambda))^{-1/2} \mu_N'(0) = \sigma^{-1} (\lambda(1-\lambda))^{1/2}.$$

(c) The Wilcoxon two sample statistic W_N satisfies the condition $A(L_W, 1/2)$ (use the same facts as in (b)) with

$$\mu_N(\vartheta) = -\int_{-\infty}^{\infty} H_N(t)\,dF_N(t) = -\frac{n}{N}\int_{-\infty}^{\infty} \left(\frac{n}{N} F(t) + \frac{m}{N} F(t-\vartheta)\right) dF(t),$$

where $\vartheta = \vartheta_2 - \vartheta_1$, and

$$\sigma_N^2(0) = \frac{2}{N}\,\lambda(1-\lambda) \iint_{0 \leqq s < t \leqq 1} s(1-t)\,ds\,dt = \frac{\lambda(1-\lambda)}{12\,N}.$$

Note that

$$\mu_N'(\vartheta) = \frac{nm}{N^2}\,\frac{1}{2\pi\sigma^2}\int_{-\infty}^{\infty} e^{-(t-\vartheta)^2/2\sigma^2}\, e^{-t^2/2\sigma^2}\,dt$$

and therefore

$$\mu_N'(0) = \frac{nm}{N^2}\,\frac{1}{2\pi\sigma^2}\int_{-\infty}^{\infty} e^{-t^2/\sigma^2}\,dt = \frac{n\,m}{2N^2\sqrt{\pi\sigma^2}}.$$

It follows that

$$L_W = \lim \frac{\mu_N'(0)}{N^{1/2}\sigma_N(0)} = \sigma^{-1}\,(\lambda(1-\lambda))^{1/2}\,(3/\pi)^{1/2}.$$

We have computed the asymptotic relative Pitman efficiencies (use Theorem 4.1.3) :

$e_P(T,S) = 1$ (t-test against van der Waerden)

$e_P(T,W) = 3/\pi$ (t-test against Wilcoxon)

$e_P(S,W) = 3/\pi$ (van der Waerden against Wilcoxon).

Concluding this section, we shall give one more result showing the relation between the ARE for estimators and the Pitman efficiency for test statistics. In particular, these efficiencies are the same for signed rank statistics and R - estimators (Recall the details from section 3.5 combined with the results of this section).

Theorem 4.1.4:

Let $\{F_\vartheta : \vartheta \in \mathbb{R}\}$ be a location model where $F=F_0$ is continuous. Let $T = \{T_n : n \geqq 1\}$ and $S = \{S_n : n \geqq 1\}$ be two sequences of statistics satisfying (3.5.18), (3.5.19) and the assumptions of Theorem 3.5.5. Assume that the associated constants A for T and A' for S have the same sign. If the asymptotic relative Pitman efficiency $e_P(T,S)$ exists then

$$e(\hat{\vartheta},\hat{\vartheta}') = e_P(T,S),$$

where $\hat{\vartheta} = \{\hat{\vartheta}_n : n \geq 1\}$ and $\hat{\vartheta}' = \{\hat{\vartheta}'_n : n \geq 1\}$ denote the estimators corresponding to T and S respectively (cf. Proposition 3.5.1).

Proof: By definition, (4.1.3 a), and Theorem 3.5.5 (3.5.24)

$$e(\hat{\vartheta},\hat{\vartheta}') = A'^2/A^2 .$$

Let $0 < \alpha < \beta < 1$ be fixed. Let φ_n and φ'_n denote tests based on T_n and S_n (cf. (4.1.12)) such that

$$\lim E_o(\varphi_n) = \lim E_o(\varphi'_n) = \alpha.$$

Observe that for $u \in \mathbb{R}$, by assumption,

$$\lim E_{u/\sqrt{n}}(\varphi_n) = \lim F^n_{u/\sqrt{n}}\left(\frac{T_n-\mu_n(u)}{\sigma_n} > \frac{k_n-\mu_n(u)}{\sigma_n}\right) =$$

$$= (2\pi)^{-1/2} \int_{\Phi^{-1}(1-\alpha)-Au}^{\infty} \exp(-t^2/2)\, dt$$

and similarly for $u' \in \mathbb{R}$

$$\lim E_{u'/\sqrt{n}}(\varphi'_n) = (2\pi)^{-1/2} \int_{\Phi^{-1}(1-\alpha)-A'u'}^{\infty} \exp(-\frac{t^2}{2})dt.$$

Choosing u and u' in such a way that

$$Au = \Phi^{-1}(1-\alpha) - \Phi^{-1}(1-\beta) = A'u',$$

it follows that the sequence $m(n) \longrightarrow \infty$ has to be chosen to satisfy $u\, n^{-1/2} = u'\,(m(n))^{-1/2}$.

Therefore

$$\lim_{n\to\infty} \frac{n}{m(n)} = \frac{u^2}{u'^2} = \frac{A'^2}{A^2} .$$

Since by assumption the asymptotic relative Pitman efficiency exists, this is obviously sufficient to prove the theorem.

2. Contiguity of probability measures

For the computation of efficiency, compare Definition 4.1.3 (4.1.11), it is necessary in certain cases to find the asymptotic distribution of test statistics T_n under a sequence of alternatives F_{ϑ_n} where $\vartheta_n \searrow \vartheta_o$. In Example 4.1.4 this condition was verified using Theorem 3.3.3; here we want to introduce another method to prove (4.1.11) which is based on the notion of contiguity.

Example 4.2.1: We start with a very simple example illustrating the situation. Consider a location model $\{ F_\vartheta : \vartheta \in \Theta \}$ and let $H_o = \{0\}$. Assume that $F = F_o$ has a density f and denote by F^n, F_ϑ^n, F^∞, F_ϑ^∞ the product measures on $\mathbb{R}^n$ resp. $\mathbb{R}^\infty$ of F and F_ϑ.

Let $A_n \subset \mathbb{R}^n$ be a sequence of measurable sets satisfying $\lim F^n(A_n) = 0$. We want to find conditions such that

$$(4.2.1) \qquad \lim_{n\to\infty} F^n_{\vartheta_n}(A_n) = \int \ldots \int_{A_n} \prod_{i=1}^n f(x_i - \vartheta_n)\, dx_1 \ldots dx_n = 0$$

holds as well.

Assume for a moment, that $f(x) = e^{-x}$. Then $\prod_{i=1}^n f(x_i - \vartheta_n) =$

$= \exp(n\vartheta_n) \prod_{i=1}^n f(x_i)$ and therefore (4.2.1) holds if

$\sup n\, \vartheta_n < \infty$. It also follows that a sequence T_n is asymptotically normal under $F^n_{\vartheta_n}$ if it is asymptotically normal under F^n, provided we assume in addition that $\lim n\, \vartheta_n$ exists.

More generally let us assume that

$$\prod_{i=1}^n f(x_i - \vartheta_n)/f(x_i) \leq g(x_1, x_2, \ldots) \quad F^\infty \text{ a.s.}$$

and that $g \in L_1(F^\infty)$ where we consider $\prod_{i=1}^n f(x_i - \vartheta_n)/f(x_i)$ to be a function on $\mathbb{R}^\infty$. Then (4.2.1) holds.

This example leads to

Definition 4.2.1: Let P_n and Q_n $(n \geq 1)$ be two sequences of probability measures P_n and Q_n on $\mathbb{R}^n$. $\{ Q_n : n \geq 1 \}$ is said to be *contiguous* to $\{ P_n : n \geq 1 \}$ if for every sequence $\{ A_n : n \geq 1 \}$ of measurable sets $A_n \subset \mathbb{R}^n$, $\lim_{n\to\infty} P_n(A_n) = 0$ implies $\lim_{n\to\infty} Q_n(A_n) = 0$.

Example 4.2.1 indicates that conditions on the likelihood quotient may force a sequence to be contiguous to another one. If two sequences $\{ P_n : n \geq 1 \}$ and $\{ Q_n : n \geq 1 \}$ are given, denote by $p_n(x)$ and $q_n(x)$ $(x \in \mathbb{R}^n)$ their densities with respect to some dominating measure μ_n $(n \geq 1)$.

Define the (usual) likelihood quotient by

$$L_n(x) = \begin{cases} q_n(x)/p_n(x) & \text{if } p_n(x) > 0,\ x \in \mathbb{R}^n \\ 1 & \text{if } p_n(x) = q_n(x) = 0,\ x \in \mathbb{R}^n \\ \infty & \text{if } p_n(x) = 0 < q_n(x),\ x \in \mathbb{R}^n. \end{cases}$$

Lemma 4.2.1: (Le Cam's first lemma) Assume that L_n converges weakly under P_n to some d.f. G satisfying $\int_0^\infty x\, dG(x) = 1$. Then $\{ Q_n : n \geq 1 \}$ is contiguous to $\{ P_n : n \geq 1 \}$.

Proof: Let $A_n \subset \mathbb{R}^n$ be measurable sets satisfying $\lim P_n(A_n) = 0$. Then for $y \in \mathbb{R}$

$$Q_n(A_n) = Q_n(A_n \cap \{L_n \leq y\}) + Q_n(A_n \cap \{L_n > y\}) =$$

$$= \int_{A_n \cap \{L_n \leq y\}} L_n \, dP_n + Q_n(A_n \cap \{L_n > y\}) \leq$$

$$\leq y\, P_n(A_n) + Q_n(\{L_n > y\}) =$$

$$= y\, P_n(A_n) + 1 - Q_n(\{L_n \leq y\}) =$$

$$= y\, P_n(A_n) + 1 - \int_{\{L_n \leq y\}} L_n \, dP_n$$

$$\xrightarrow[n \to \infty]{} 1 - \int_0^y x\, dG(x) \xrightarrow[y \to \infty]{} 0 \quad \text{(by assumption).}$$

The proof follows immediately from this computation.

Corollary: If in the preceding lemma G is log-normal $(-\frac{1}{2}\sigma^2, \sigma^2)$, then $\{ Q_n : n \geq 1 \}$ is contiguous to $\{P_n : n \geq 1\}$.

Proof: Recall that a variable is called log-normal (a, σ^2), if its logarithm is $N(a, \sigma^2)$. Thus, if the d.f. of Y is G as in the corollary, then

$$\int_0^\infty x\, dG(x) = EY = (2\pi\sigma^2)^{-1/2} \int \exp\left[x - \frac{(x+\sigma^2/2)^2}{2\sigma^2}\right] dx = 1.$$

The other goal described in the beginning of this section is to establish the weak convergence under contiguous alternatives. In view of Lemma 4.2.1 one can do this using Le Cam's third lemma.

Lemma 4.2.2: Let the sequences $\{ Q_n : n \geq 1 \}$ and $\{ P_n : n \geq 1 \}$ be given. Assume that $\{ T_n : n \geq 1 \}$ is a sequence of statistics $T_n : \mathbb{R}^n \longrightarrow \mathbb{R}$ such that the two-dimensional vectors $(T_n, \log L_n)$ $(n \geq 1)$ converge weakly under P_n to a (two-dimensional) normal distribution N(,) with expectation $(\mu_1, -1/2\, \sigma_2^2)$ and covariance operator

(4.2.2) $$\Sigma = \begin{pmatrix} \sigma_1^2 & \sigma_{12} \\ \sigma_{12} & \sigma_2^2 \end{pmatrix} .$$

Then T_n ($n \geq 1$) converges weakly under Q_n to $N(\mu_1+\sigma_{12}, \sigma_1^2)$.

<u>Proof:</u> The assumption implies in particular that $\log L_n$ converges weakly to $N(-\sigma_2^2/2, \sigma_2^2)$. We can apply Lemma 4.2.1 and its corollary to conclude that $\{ Q_n : n \geq 1 \}$ is contiguous to $\{ P_n : n \geq 1 \}$.

Therefore, for every sequence $c_n \longrightarrow \infty$,

(4.2.3) $$\lim_{n\to\infty} Q_n(|\log L_n| \geq c_n) = 0.$$

In order to prove the lemma, observe first that for $x \in \mathbb{R}$

$$Q_n(T_n \leq x) = Q_n(T_n \leq x, |\log L_n| \leq c_n) + Q_n(T_n \leq x, |\log L_n| > c_n) =$$

$$= \int 1_{\{T_n \leq x, |\log L_n| \leq c_n\}} \exp(\log L_n)\, dP_n + Q_n(T_n \leq x, |\log L_n| > c_n).$$

In view of (4.2.3), $Q_n(T_n \leq x)$ has the same limit as

$$\int 1_{\{T_n \leq x, |\log L_n| \leq c_n\}} L_n\, dP_n,$$

provided $c_n \longrightarrow \infty$ (no matter how slowly).

By assumption, for fixed $c = c_n$,

(4.2.4) $$\lim_{n\to\infty} \int 1_{\{T_n \leq x, |\log L_n| \leq c\}} L_n\, dP_n = \int 1_{\{u \leq x, v \leq c\}} \exp v\, dN(u,v).$$

It is easy to construct a sequence $c_n \longrightarrow \infty$ (slowly) such that (4.2.4) still holds true with $c = c_n$ on the left-hand side and $c = \infty$ on the right-hand side. Therefore

$$\lim_{n\to\infty} Q_n(T_n \leq x) = \int_{-\infty}^{x} \int_{-\infty}^{\infty} \exp v \; dN(u,v) =$$

$$= (2\pi\sigma_1\sigma_2)^{-1}(1-\rho^2)^{-1/2} \iint 1_{(-\infty,x]}(u) \exp v \; \exp - \frac{(u-\mu_1)^2\sigma_1^{-2} - 2\rho/\sigma_1\sigma_2\, (v+\sigma_2^2/2)(u-\mu_1) + (v+\sigma_2^2/2)^2\sigma_2^{-2}}{2(1-\rho^2)} \,dudv$$

$$= \sigma_1^{-1} (2\pi)^{-1/2} \int_{-\infty}^{x} \exp\left(- \frac{(u-\mu_1-\sigma_{12})^2}{2\sigma_1^2} \right) du,$$

where $\rho = \sigma_{12} \sigma_1^{-1} \sigma_2^{-1}$. (Exercise: Check the details of the proof.)

We shall now indicate by an example, how Lemmas 4.2.1 and 4.2.2 can be used to prove asymptotic normality under contiguous alternatives.

Example 4.2.2: (including Lemmas 4.2.3 - 4.2.6 and Proposition 4.2.1)

Let $\{ F_\vartheta : \vartheta \in \Theta\}$ be a location model (with $F_\vartheta(x) = F(x-\vartheta)$). Assume that F has a C^1-density f such that the Fisher information I(F) is finite. Let $X_1, X_2, \ldots$ be independent, F-distributed random variables.

Lemma 4.2.3: We have

$$(4.2.5) \qquad \lim_{n\to\infty} E\left| n\left\{ \frac{f(X_1-\vartheta_n)^{1/2}}{f(X_1)^{1/2}} - 1 \right\}^2 - \frac{I(F)}{4}\right| = 0$$

and

$$(4.2.6) \qquad \lim_{n\to\infty} n\, E\left(\left\{ \frac{f(X_1-\vartheta_n)}{f(X_1)} \right\}^{1/2} - 1\right) = - \frac{I(F)}{8},$$

provided

$$(4.2.7) \qquad \lim_{n\to\infty} n\, \vartheta_n^2 = 1.$$

Proof: Since by Fubini's theorem

$$\int_{-\infty}^{\infty} \left\{ \frac{f(x-\vartheta_n)^{1/2}}{f(x)^{1/2}} - 1 \right\}^2 f(x)\, dx =$$

$$= \vartheta_n^2 \int_{-\infty}^{\infty} \left\{ \frac{f(x-\vartheta_n)^{1/2} - f(x)^{1/2}}{\vartheta_n} \right\}^2 dx =$$

$$= \vartheta_n^2 \int_{-\infty}^{\infty} \left(\vartheta_n^{-1} \int_0^{\vartheta_n} \frac{f'(x-t)}{2f(x-t)^{1/2}}\, dt \right)^2 dx \leq$$

$$\leq \vartheta_n^2 \int_{-\infty}^{\infty} \vartheta_n^{-1} \int_0^{\vartheta_n} \frac{f'^2(x-t)}{4\, f(x-t)}\, dt\, dx =$$

$$= \vartheta_n^2 (4\vartheta_n)^{-1} \int_0^{\vartheta_n} \int_{-\infty}^{\infty} \frac{f'^2(x-t)}{f(x-t)}\, dx\, dt = \frac{\vartheta_n^2\, I(F)}{4},$$

and since

$$\vartheta_n^{-1}\left(f(x-\vartheta_n)^{1/2} - f(x)^{1/2} \right) \longrightarrow -f'(x)/2f(x)^{1/2},$$

it follows from Fatou's lemma and (4.2.7) that

$$\text{(4.2.8)} \qquad \lim_{n\to\infty} n \int_{-\infty}^{\infty} \left\{ \frac{f(x-\vartheta_n)^{1/2}}{f(x)^{1/2}} - 1 \right\}^2 f(x)\, dx = \frac{I(F)}{4}.$$

Moreover, for every measurable set A, we have

$$\text{(4.2.9)} \qquad \lim_{n\to\infty} n \int_A \left\{ \frac{f(x-\vartheta_n)^{1/2}}{f(x)^{1/2}} - 1 \right\}^2 f(x)\, dx = \int_A \frac{f'^2(x)}{4\, f(x)} dx$$

(4.2.9) holds, because otherwise there exists a subsequence n_k $(k \geq 1)$ such that

$$\lim_{k\to\infty} n_k \int_A \left\{ \frac{f(x-\vartheta_{n_k})^{1/2}}{f(x)^{1/2}} - 1 \right\}^2 f(x)dx > \int_A \frac{f'^2(x)}{4\, f(x)} dx .$$

Hence by Fatou's lemma

$$\lim_{k\to\infty} n_k \left[\int_A \left\{ \frac{f(x-\vartheta_{n_k})^{1/2}}{f(x)^{1/2}} - 1 \right\}^2 f(x)\, dx + \right.$$

$$\left. + \int_{A^c} \left\{ \frac{f(x-\vartheta_{n_k})^{1/2}}{f(x)^{1/2}} - 1 \right\}^2 f(x)\, dx \right] > \frac{I(F)}{4}$$

contradicting the above equality (4.2.8). Observe that (4.2.9) implies (4.2.5).

The proof of (4.2.6) follows now easily. Note that

$$\int_{-\infty}^{\infty} \left(f(x-\vartheta_n)^{1/2} - f(x)^{1/2} \right)^2 dx =$$

$$= 2 - 2 \int_{-\infty}^{\infty} \left(\frac{f(x-\vartheta_n)}{f(x)} \right)^{1/2} f(x)\, dx.$$

Therefore

$$\lim_{n\to\infty} n\, E\left\{ \left(\frac{f(X_1-\vartheta_n)}{f(X_1)} \right)^{1/2} - 1 \right\} =$$

$$= - \lim_{n\to\infty} 1/2 \int_{-\infty}^{\infty} \left\{ \frac{f(x-\vartheta_n)^{1/2} - f(x)^{1/2}}{\vartheta_n} \right\}^2 dx =$$

$$= - \frac{I(F)}{8}$$

by (4.2.8).

<u>Lemma 4.2.4:</u> We have

$$\lim_{n\to\infty} nP(|f(X_1-\vartheta_n)^{1/2} f(X_1)^{-1/2} - 1| > \delta) = 0$$

for every $\delta > 0$ where ϑ_n satisfies (4.2.7).

Proof:

$$P\left(\left| \left\{ \frac{f(X_1-\vartheta_n)}{f(X_1)} \right\}^{1/2} - 1 \right| \geq \delta \right) =$$

$$= P\left(\left(\left\{ \frac{f(X_1-\vartheta_n)}{f(X_1)} \right\}^{1/2} - 1 \right)^2 - \frac{I(F)}{4n} \geq \delta^2 - \frac{I(F)}{4n} \right)$$

If n is so large that $I(F) < 2n\delta^2$, then by Lemma 4.2.3 (4.2.5) and the Markov inequality

$$n\, P\left(\left| \left\{ \frac{f(X_1-\vartheta_n)}{f(X_1)} \right\}^{1/2} - 1 \right| \geq \delta \right) \leq$$

$$\leq 2\,\delta^{-2}\, E \left| n\left(\left\{ \frac{f(X_1-\vartheta_n)}{f(X_1)} \right\}^2 - 1 \right)^2 - \frac{I(F)}{4} \right| \longrightarrow 0.$$

The next lemma is in fact a version of Le Cam's second lemma.

Lemma 4.2.5: If (4.2.7) holds then

$$\lim_{n\to\infty} \left(\log L_n - 2 \sum_{i=1}^{n} \left(\left\{ \frac{f(X_i-\vartheta_n)}{f(X_i)} \right\}^{1/2} - 1 \right) + \frac{I(F)}{4} \right) = 0$$

in probability.

Proof: Using Taylor's theorem for log (1+z) we find

$$\log \frac{f(X_i-\vartheta_n)}{f(X_i)} = 2 \log \left(1 + \left(\left\{ \frac{f(X_i-\vartheta_n)}{f(X_i)} \right\}^{1/2} - 1 \right) \right) =$$

$$= W_i - W_i^2/4 \int_0^1 \frac{2(1-t)}{(1+tW_i/2)^2}\, dt$$

where $W_i = 2\, f(X_i-\vartheta_n)^{1/2}\, f(X_i)^{-1/2} - 2$ (omitting the index n in W_i).

Thus

$$\log L_n = \sum_{i=1}^{n} \left[W_i - W_i^2/4 + W_i^2/4 \left(1 - \int_0^1 \frac{2(1-t)}{(1+tW_i/2)^2}\, dt \right) \right]$$

Let us first show that the last term tends to zero in probability as $n \longrightarrow \infty$.

By Lemma 4.2.4, $\max \{W_i : 1 \leq i \leq n\} \longrightarrow 0$ in probability, hence, given $\varepsilon > 0$, there exists an $n_0 \in \mathbb{N}$ such that

$$\left| 1 - \int_0^1 \frac{2(1-t)}{(1+tW_i/2)^2}\, dt \right| < \varepsilon \qquad (1 \leq i \leq n)$$

with probability $\geq 1 - \varepsilon$, if $n \geq n_0$. If $n \geq n_0$, it follows that with probability $\geq 1 - \varepsilon$

$$\left| \sum_{i=1}^{n} W_i^2 \left(1 - \int_0^1 \frac{2(1-t)}{(1+tW_i/2)^2}\, dt \right)\right| \leq \varepsilon \sum_{i=1}^{n} W_i^2$$

and in view of Lemma 4.2.3 (4.2.5)

$$\sum_{i=1}^{n} W_i^2 \longrightarrow I(F) \quad \text{in probability.}$$

This proves our first claim.

Observe now that the lemma follows immediately using (4.2.5) again.

Lemma 4.2.6: Let W_i be as in the proof of Lemma 4.2.5 and define $V_i = -\vartheta_n f'(X_i)/f(X_i)$ $(1 \leq i \leq n$; omitting the index n again). Then under the condition (4.2.7)

$$\lim_{n\to\infty} \operatorname{Var}\left(\sum_{i=1}^{n} W_i - V_i\right) = 0.$$

Proof: Since

$$\operatorname{Var}\left(\sum_{i=1}^{n} W_i - V_i\right) = n \operatorname{Var}(W_1 - V_1)$$

the lemma will follow, if we can show that $\vartheta_n^{-2} E(W_1 - V_1)^2 \longrightarrow 0$ (in this case $\lim n\, (EW_1)^2 = 0$). Observe that

$$\vartheta_n^{-2} EV_1^2 = I(F) \quad \text{and that} \quad \lim \vartheta_n^{-2} EW_1^2 = I(F)$$

by Lemma 4.2.3. Moreover,

$$\limsup_{n\longrightarrow\infty} \vartheta_n^{-2} E|W_1V_1| \leq \limsup_{n\longrightarrow\infty} (\vartheta_n^{-4} EW_1^2\, EV_1^2)^{1/2} = I(F)$$

and using $\vartheta_n^{-1} W_1 \longrightarrow \vartheta_n^{-1} V_1$ and Fatou's lemma, it follows that

$$\lim_{n\to\infty} \vartheta_n^{-2} E|W_1V_1| = I(F).$$

Similar to the arguement in the proof of Lemma 4.2.3 one can show that

$$\lim_{n\to\infty} \vartheta_n^{-2} EW_1V_1 = I(F).$$

Putting everything together

$$\lim_{n\to\infty} \vartheta_n^{-2} E(W_1-V_1)^2 = 0.$$

Proposition 4.2.1: Under the general assumptions and (4.2.7) we have

$$\text{(4.2.10)} \quad \log L_n - \sum_{i=1}^{n} V_i + \frac{I(F)}{2} \longrightarrow 0 \quad \text{in probability,}$$

(4.2.11) $\sum_{i=1}^{n} V_i$ converges weakly to $N(0,I(F))$,

(4.2.12) $\log L_n$ converges weakly to $N(-I(F)/2,\ I(F))$, consequently $\{ F^n_{\vartheta_n} : n \geq 1 \}$ is contiguous to $\{ F^n : n \geq 1 \}$.

Proof: (4.2.10) follows immediately from Lemmas 4.2.3, (4.2.6), 4.2.5 and 4.2.6. (4.2.11) is a consequence of the central limit theorem for i.i.d. random variables and (4.2.12) follows from (4.2.10) and (4.2.11). The contiguity is Le Cam's first lemma (corollary to Lemma 4.2.1).

Example 4.2.3: Let $\{ F_\vartheta : \vartheta \in \Theta \}$ be a location model as in Example 4.2.2, where $F = F_0$ has the density f with finite Fisher information I(F). We assume in addition that (4.2.7) holds for the alternatives ϑ_n (i.e. $n\vartheta_n \longrightarrow 1$). Let us now consider a sequence T^*_n $(n \geq 1)$ of simple linear rank statistics

$$T^*_n = \sum_{i=1}^{n} c_{in}\ h(R_{in}/n+1)$$

where R_{in} denotes the rank of X_i among the independent F-distributed random variables $X_1,\dots,X_n$. We shall make the assumptions of Theorem 3.4.1, in particular we may assume that $\int h(t)\ dt = 0$ and $0 < \int h^2(t)\ dt < \infty$. Setting

$$D_n = \left[\sum_{i=1}^{n} (c_{in} - \bar{c}_n)^2 \right]^{-1/2} \qquad (\bar{c}_n = n^{-1} \sum_{i=1}^{n} c_{in}),$$

$a_n(i) = h(i/(n+1))$, $\bar{a}_n = n^{-1} \sum a_n(i)$ and $T_n = T^*_n - n\bar{a}_n\bar{c}_n$

it has been shown in Theorem 3.4.1 that $n^{-1/2}D_nT_n$ converges weakly to $N(0,\sigma^2)$ where $\sigma^2 = \int h^2(t)\ dt$. (Recall the details. The corollary to Theorem 3.4.3 (p.135) shows that we can replace T_n by T^*_n in certain cases.)

In order to show the asymptotic normality of these statistics under the contiguous alternatives $F^n_{\vartheta_n}$, we shall apply Lemma 4.2.2.

We have to show that $(n^{-1/2}D_nT_n,\ \log L_n)$ is asymptotically normal and, moreover, we have to determine the limiting expectation (μ_1,μ_2) and the limiting covariance matrix (4.2.2).

Clearly $\mu_1 = 0$ and by Proposition 4.2.1 $\mu_2 = -\ I(F)/2$. Also it is obvious that $\sigma_1^2 = \int h^2(t)\ dt$ and $\sigma_2^2 = I(F)$.

From Proposition 4.2.1,(4.2.10), it is clear that $\log L_n$ is asymptotically equivalent to $\sum V_i - I(F)/2$ with $V_i = -\vartheta_n f'(X_i)/f(X_i)$ $(1 \leq i \leq n)$. Therefore, $(n^{-1/2} D_n T_n, \log L_n)$ may be replaced by

$$(4.2.13) \quad \left(n^{-1/2} D_n T_n, \ -\vartheta_n \sum_{i=1}^{n} f'(X_i)/f(X_i) - I(F)/2\right).$$

It follows from (4.2.11) that the second coordinate converges weakly to $N(-I(F)/2, I(F))$, and since both coordinates are independent, the vectors in (4.2.13) are asymptotically normal. In fact, the first coordinate is a function of the ranks while the second one can be written as a function of the order statistic. (Exercise: Show that the rank vector $R = (R_1, \ldots, R_n)$ is independent of the order statistic $X^{(n)} = (X_{(1)}, \ldots, X_{(n)})$ under i.i.d. random variables $X_1, \ldots, X_n$ with continuous d.f.) Thus by Lemma 4.2.2, $n^{-1/2} D_n T_n$ converges weakly to $N(0, \sigma^2)$ w.r.to $F^n_{\vartheta_n}$ where σ^2 is as before.

Let us now consider a two-sample problem, i.e. we want to show the asymptotic normality of $n^{-1/2} D_n T_n$ under the contiguous alternatives $Q_n = F_0^p \times F^q_{\vartheta_n}$ where ϑ_n satisfies (4.2.7) and where $\lim p/n = \lambda$. Here $C_{in}^n = 1$ if $i \leq p$ and $= 0$ if $i > p$ (consequently $D_n^2 = n^2/pq$) and

$$L_n(x_1, \ldots, x_n) = \prod_{i=p+1}^{n} f(x_i - \vartheta_n)/f(x_i) \ .$$

Check the following details as an exercise: Lemmas 4.2.3 and 4.2.4 remain valid whereas Lemma 4.2.5 reads

$$\lim \ \log L_n - 2\left(\sum_{i=p+1}^{n} \left[\frac{f(X_i - \vartheta_n)}{f(X_i)}\right]^{1/2} - 1\right) + \frac{qI(F)}{4n} = 0.$$

Also, Lemma 4.2.6 is true in the form $\operatorname{Var}\left(\sum_{i>p} W_i - V_i\right) \longrightarrow 0$.

Thus Proposition 4.2.1 has the following form:

$$(4.2.10)' \quad \log L_n - \sum_{i=p+1}^{n} V_i + \frac{qI(F)}{2n} \longrightarrow 0 \quad \text{in probability}$$

$$(4.2.11)' \quad \sum_{i=p+1}^{n} V_i \longrightarrow N(0, (1-\lambda) I(F))$$

$$(4.2.12)' \quad \log L_n \longrightarrow N(-(1-\lambda) I(F)/2, (1-\lambda) I(F)) \ .$$

It follows that $(n^{-1/2} D_n T_n, \log L_n)$ is asymptotically equivalent to

$$(4.2.14) \quad \left(n^{-1/2} D_n T_n, \ -\vartheta_n \sum_{i=p+1}^{n} f'(X_i)/f(X_i) - \frac{qI(F)}{2n}\right).$$

Observe that - in the present situation - the second coordinate is not a function of the order statistic. Let $h_n(t) = a_n(i)$ if $i/(n+1) \leq t < (i+1)/(n+1)$ (cf. p.117). The vector $(a_n(R_1),\dots,a_n(R_n),h_n(F(X_1)),\dots,h_n(F(X_n)))$ has the distribution defined in (3.1.6) (Exercise). Therefore, part (B) of the proof of Theorem 3.1.1 shows that

$$n^{-1/2} D_n \left(T_n - \sum_{i=1}^{n} (c_{in}-\bar{c}_n)(h_n(F(X_i)) - \bar{a}_n)\right) \longrightarrow 0 \quad \text{in } L_2$$

and hence (4.2.14) is asymptotically equivalent to the statistics

$$(4.2.15) \qquad \left(\sum_{i=1}^{n} n^{-1/2} D_n (c_{in}-\bar{c}_n)(h_n(F(X_i))-\bar{a}_n), -\vartheta_n \sum_{j=p+1}^{n} \frac{f'(X_j)}{f(X_j)} - \frac{qI(F)}{2n} \right).$$

It is easy to see that these vectors converge weakly to a normal distribution (using the Cramér-Wold device, for example). With the notation as before we have $\mu_1 = 0$, $\mu_2 = -(1-\lambda)I(F)/2$, $\sigma_1^2 = \sigma^2 = \int h^2(t)\,dt$, $\sigma_2^2 = (1-\lambda)I(F)$ and, in view of (4.2.7),

$$\sigma_{12} = \lim_{n\to\infty} - \sum_{j=p+1}^{n} n^{-1} D_n (c_{jn}-\bar{c}_n) \, E\, h_n(F(X_j)) \frac{f'(X_j)}{f(X_j)} =$$

$$= (\lambda(1-\lambda))^{1/2} \lim_{n\to\infty} \int h_n(t) \frac{f'(F^{-1}(t))}{f(F^{-1}(t))} dt =$$

$$= (\lambda(1-\lambda))^{1/2} \int h(t) \frac{f'(F^{-1}(t))}{f(F^{-1}(t))} dt.$$

It follows now from Lemma 4.2.2 that $n^{-1/2} D_n T_n$ is asymptotically normal $N(\sigma_{12},\sigma^2)$, where σ^2 is as before.
(Exercise: Check the details of the preceding arguments. Use also Theorem 3.4.3 (and partial integration) to show the same result, though only for special score functions.)

Example 4.2.4: Let $\{ F_\vartheta : \vartheta \in \Theta \}$ be a location model as in Examples **4.2.2 and** 4.2.3. Let us now consider U-statistics $U_n(h)$ where the kernel h has degree m, is non-degenerate and satisfies $\int\dots\int h^2(x_1,\dots,x_m)\, dF(x_1)\dots dF(x_m) < \infty$ and $\int\dots\int h(x_1,\dots,x_m)\, dF(x_1)\dots dF(x_m) = \vartheta(F) = 0$. According to Proposition 1.3.1 $n^{1/2}U_n(h)$ is L_2-equivalent to

$$m\, n^{-1/2} \sum_{i=1}^{n} \tilde{h}_1(X_i)$$

where $\tilde{h}_1(x) = \int\ldots\int h(x,x_2,\ldots,x_m)\, dF(x_2)\ldots dF(x_m)$.
Since by Proposition 4.2.1, (4.2.10), $\log L_n$ is equivalent to $\sum V_i - I(F)/2$, it follows that the pair $(n^{1/2}U_n(h), \log L_n)$ is asymptotically normal with expectation $(0, -I(F)/2)$ and limiting covariance matrix Σ defined by

$$\sigma_1^2 = m^2\,\zeta_1 \qquad (\zeta_1 = \mathrm{Var}_F(\tilde{h}_1))$$

$$\sigma_2^2 = I(F) \quad \text{and}$$

$$\sigma_{12} = \mathrm{Cov}\,(m\,\tilde{h}_1(X_1),\ f'(X_1)/f(X_1)) =$$

$$= m\int \tilde{h}_1(x)\, f'(x)\, dx \qquad (\text{cf. } (4.2.2)).$$

We can apply Lemma 4.2.2 again to conclude that $n^{1/2}\, U_n(h)$ is asymptotically normal under $F^n_{\vartheta_n}$ with expectation σ_{12} and variance σ_1^2.

Notes on chapter 4:

The ARE usually is defined by (4.1.3a) and contained in most of the books in the reference. Theorem 4.1.2 is due to Rösler and the author. The conditions in Definition 4.1.3 are a version of the Pitman condition and Theorem 4.1.3 is a form of the Pitman-Noether theorem (cf. Serfling (1980)). In section 2 on contiguity le Cam's theory is discussed briefly following in parts Hájek, Šidák (1967). More on contiguity can be found in Roussas book (1972) for example.

References

The following references are closely related to the topics contained in this book. The list, however, is by far not complete.

Akahira,M.;Takeuchi,K. (1981): Asymptotic Efficiency of Statistical Estimators. Lecture Notes in Statistics, 7, Springer, New York.

Berk,R.H. (1966): Limiting behavior of posterior distributions when the model is incorrect. Ann. Math. Statist., 37,51-58

Billingsley,P. (1968): Convergence of Probability Measures. Wiley, New York

Cambanis,S.;Simons,G.;Stout,W. (1976): Inequalities for Ek(X,Y) when the marginals are fixed. Z. Wahrscheinlichkeitsth., 36, 285-294

Dehling,H.;Denker,M.;Philipp,W. (1983): Invariance principles for von Mises and U-statistics. to appear Z. Wahrscheinlichkeitsth.

Denker,M.;Grillenberger,C.;Keller,G. (1981): A note on invariance principles for v. Mises' statistics. to appear Metrika

Denker,M.;Puri,M.L.;Rösler,U. (1982): A sharpening of the remainder term in the higher-dimensional central limit theorem for multilinear rank statistics. to appear J. Multivar. Analysis.

Denker,M.;Rösler,U. (1981): Some contributions to Chernoff-Savage theorems. to appear Statistics & Decision

Denker,M.;Rösler,U. (1982): A note on the asymptotic normality of rank statistics. to appear Proceedings Conference on Probability Theory, Veszprém, Hungary 1982

Doob,J.L. (1953): Stochastic Processes. Wiley, New York

Feller,W. (1950): An Introduction to Probability Theory and Its Applications. Vol. 1. Wiley, New York

Fernholz,L.T. (1983): Von Mises Calculus for Statistical Functionals. Lecture Notes in Statistics, 19, Springer, New York

Filippova,A.A. (1962): Mises' theorem on the asymptotic behavior of functionals of empirical distribution function and its statistical applications. Theory Prob. Appl., 7, 24-57

Gänssler,P.;Stute,W. (1977): Wahrscheinlichkeitstheorie. Springer, Berlin-Heidelberg-New York

Hájek,J. (1961): Some extensions of the Wald-Wolfowitz-Noether theorem. Ann. Math. Statist., 32, 506-523

Hájek,J. (1968): Asymptotic normality of simple linear rank statistics under alternatives. Ann. Math. Statist., 39, 325-346

Hájek,J.;Šidák,Z. (1967): Theory of Rank Tests. Academic Press, New York

Hoeffding,W. (1948): A class of statistics with asymptotically normal distribution. Ann. Math. Statist., 19, 293-325

Hoeffding,W. (1950): "Optimum" non-parametric tests. Proc. IInd Berkeley Symp. 83-92

Hodges,J.L. Jr.;Lehmann,E.L. (1963): Estimates of location based on rank tests. Ann. Math. Stat., 34, 598-611

Hollander,M.;Proschan,F. (1972): Testing whether new is better than used. Ann. Math. Statist., 43, 1136-1146

Huber,P.J. (1981): Robust Statistics. Wiley, New York

Lehmann,E.L. (1949): On the theory of some non-parametric hypotheses. Ann. Math. Statist., 20, 28-45

Lehmann,E.L. (1951): Consistency and unbiasedness of certain nonparametric tests. Ann. Math. Statist., 22, 165-179

Lehmann,E.L. (1959): Testing Statistical Hypotheses. Wiley, New York

Miller,R.G. Jr.;Sen,P.K. (1972): Weak convergence of U-statistics and von Mises' differentiable statistical functions. Ann. Math. Statist., 43, 31-41

Puri,M.L.,Sen,P.K. (1971): Nonparametric Methods in Multivariate Analysis. Wiley, New York

Randles,R.H.;Wolfe,D.A. (1979): Introduction to The Theory of Nonparametric Statistics. Wiley, New York

Roussas,G.G. (1972): Contiguity of Probability Measures: Some Applications in Statistics. Cambridge University Press, Cambridge

Sen,P.K. (1974): Almost sure behavior of U-statistics and von Mises' differentiable statistical functions. Ann.

Statist., 2, 387-395

Sen,P.K. (1974): Weak convergence of generalized U-statistics. Ann. Prob., 2, 90-102

Serfling,R.J. (1971):The law of iterated logarithm for U-statistics and related von Mises' statistics. Ann. Math. Statist., 42, 1794

Serfling,R.J. (1980): Approximation Theorems of Mathematical Statistics. Wiley, New York

von Mises,R. (1947): On the asymptotic distribution of differentiable statistical functions. Ann. Math. Statist., 18, 309-348

Subject Index